Terence Etchells, Mark Hunter, Jo
Stefano Pozzi, Andrew Rothery

Mathematical Activities with Computer Algebra

a photocopiable resource book

British Libary Catalogouing in Publication Data
A catalogue record for this book is available from the British Library

ISBN 0-86238-405-2

Printed in Sweden
Studentlitteratur
ISBN 91-44-61681-3

Printing:	1 2 3 4 5 6 7 8 9 10	2001 00 99 98 97

Contents

Preface 5
Introduction 7

Multiplying Straight Lines
Activity Worksheet 14
Help Sheet 15
Teaching Notes 16

Equation of a Tangent
Activity Worksheet 18
Help Sheet 19
Teaching Notes 20

Taxing Functions
Activity Worksheet 22
Help Sheet 23
Teaching Notes 24

The Tile Factory
Activity Worksheet 26
Help Sheet 27
Teaching Notes 28

Visualising Functions and Derivatives
Activity Worksheet 32
Help Sheet 33
Teaching Notes 34

The Approximate Derivative Function
Activity Worksheet 36
Help Sheet 37
Teaching Notes 38

Sketching Graphs
Activity Worksheet 40
Help Sheet 41
Teaching Notes 42

Population and Pollution
Activity Worksheet 44
Help Sheet 45
Teaching Notes 46

Max Cone
Activity Worksheet 48
Help Sheet 49
Teaching Notes 50

Optimising Transport Costs
Activity Worksheet 52
Help Sheet 53
Teaching Notes 54

The Area Under a Curve
Activity Worksheet 56
Help Sheet 57
Teaching Notes 58

Enclosed Areas
Activity Worksheet 62
Help Sheet 63
Teaching Notes 64

A Function whose Derivative is Itself
Activity Worksheet 66
Help Sheet 67
Teaching Notes 68

Wine Glass Design
Activity Worksheet 70
Help Sheet 71
Teaching Notes 72

The Limit of a Sequence
Activity Worksheet 74
Help Sheet 75
Teaching Notes 76

Visualising Taylor Approximations
Activity Worksheet 78
Help Sheet 79
Teaching Notes 80

Visualising Matrix Transformation
Activity Worksheet 82
Help Sheet 83
Teaching Notes 84

Blood Groups
Activity Worksheet 86
Help Sheet 87
Teaching Notes 88

Circular Motion
Activity Worksheet 92
Help Sheet 93
Teaching Notes 94

Swing Safety
Activity Worksheet 98
Help Sheet 99
Teaching Notes 100

No Turning Back
Activity Worksheet 104
Help Sheet 105
Teaching Notes 106

Modelling the Sine Function
Activity Worksheet 110
Help Sheet 111
Teaching Notes 112

Solving Equations with Tangents
Activity Worksheet 116
Help Sheet 117
Teaching Notes 118

Appendices 121
Derive 123
MathPlus/Theorist 124
Maple 125
TI-92 126
Mathematica 127
Macsyma 128

Preface

This book of activities arose from a belief, founded in practice, that there are many ways that computer algebra systems can enhance the learning of mathematics and that some 'starters' in the form of worksheets will help teachers appreciate some of these ways.

We wish to stress that this is a book of activities rather than a textbook. This reflects our experience that, at the time of writing, most teachers are experimenting with using computer algebra with their students. We have not tried to be comprehensive but have merely chosen some activities that we think will be successful – in the hope that these ideas will spur you on to develop your own worksheets. We have, however, tried to vary the content and the mathematical level of difficulty.

The focus of this book is the mathematics and this is partly why we have opted for writing our worksheets in a generic style – to avoid "press-this-press-that" worksheets where there is an over-emphasis on the technology. Another reason for writing in a generic style is that computer algebra systems change quickly and new systems appear . We have assumed that you know the basics of how to use at least one such system and that you have introduced these basics to your students.

Original ideas are few and far between. We have obtained ideas from many sources – not always consciously. We would like to express particular thanks to Carl Leinbach for his excellent book "Calculus Laboratories Using DERIVE", and the authors of the Association of Teachers of Mathematics (UK) "Algebra at A-level – how the curriculum might change with computer algebra" and "These have worked for us at A-level" series. We would also like to thank the many people who helped us with revisions and redrafts – with particular thanks to Philip Yorke and Sten Olsson of Chartwell-Bratt for all their work.

Writing the book was great fun and we hope you enjoy it. We would like to hear your comments on how we could improve the materials. Or if you have any ideas for new worksheets, contact our publisher and, who knows, you could be a co-author of any sequels to this book.

Andrew, Stefano, John, Mark and Terence

Introduction

The aim of the book

Computer algebra systems will have a significant impact on the way people carry out their mathematics. The ability to work out algebraic manipulation and calculus using a machine will have a dramatic effect on the mathematics curriculum. Indeed we feel that the effect will be more than that of calculators, spreadsheets and graphic calculators. Originally developed for use by mathematics professionals on large computers, computer algebra systems are now widely available for students, either as software packages on personal computers or built into hand-held machines. Though computer algebra systems have been in existence for some years, the recent availability of this technology at a personal level will lead to a sudden increase in its use.

This book is aimed at students in the over-16 age range in sixth forms, colleges and first year higher education in universities in the UK and USA. We provide a starting point for the teacher or lecturer who wishes to introduce computer algebra activities into their teaching approach – the book covers a range of different topics at different levels of ability. The design of the book enables it to be used flexibly for class, lab, group or individual work.

We do not link our book to any particular computer algebra software – we believe that its emphasis should be on mathematics. Provided your students have had a basic introduction to the computer algebra system you use, they will be able to follow the materials. We use language common to all computer algebra systems ('keywords') when we give hints and advice. A further, practical reason for publishing material which is independent of software is that there is presently a wide range of different packages on the market, each with quite different user interfaces. Additionally, most packages are continually being updated, and new versions of the same program can be substantially different from the earlier ones. Furthermore, your students may have a different program at home or on their hand held machine from the one in use at school, college or university. In view of this diversity, the materials presented here focus on mathematical ideas and problems which form a basis for the use of computer algebra.

Information and instructions on commonly available computer algebra systems is given in the appendices and there are explanations of how each implements the keywords of computer algebra language.

The choice of topics is intended to be compatible with A Level mathematics in the UK, SCAA guidelines and NCTM standards in the USA. The activities have a mathematical focus and we have chosen them because they are ideally suited to working with computer algebra; indeed, they would be difficult, inconvenient and in some cases virtually impossible without the use of a computer algebra system. Together, we feel they illustrate the potential impact of this revolutionary technology and will help teachers and lecturers make a start in implementing its use.

What the book offers

- 23 photocopiable student worksheets
- 23 photocopiable optional student help sheets
- detailed teaching notes and solutions
- an approach which permits working with any computer algebra system
- a range of mathematical topics in the 16+ curriculum
- activities which easily align with standard teaching topics

Organisation and content

The activities are at different levels of difficulty, making it easier to identify activities for the particular needs of any group of students. Some activities cover mathematical techniques and concepts and many relate to an applicable mathematics or modelling context. In some cases, suggested extensions to activities provide ample material for an extended project.

Topic areas	*List of activities*
Functions, graphs and calculus	Multiplying Straight Lines
	Equation of a Tangent
	Taxing Functions
	The Tile Factory
Differentiation and Optimisation	Visualising Functions and Derivatives
	The Approximate Derivative Function
	Sketching Graphs
	Population and Pollution
	Max Cone
	Optimising Transport Costs
Integration	The Area Under a Curve
	Enclosed Areas
	A Function whose Derivative is Itself
	Wine Glass Design
Sequences and Series	The Limit of a Sequence
	Visualising Taylor Approximations
Vectors and Matrices	Visualising Matrix Transformations
	Blood Groups
Mechanics	Circular Motion
	Swing Safety
	No Turning Back
Trigonometry	Modelling the Sine Function
Numerical Methods	Solving Equations with Tangents

The choice of content provides an activity within each main area of the curriculum for the intended age group. There is a strong emphasis on calculus and graphs and a substantial number of the activities involve a modelling context.

Using the book

Activity Worksheets

The activity worksheets each provide a problem to solve or the starting point for an area of investigation. Copyright permission is given to reproduce these worksheets within one institution so that they may be used cheaply and easily. They can provide material for a class or computer lab session – equally they can be given to individual students for independent study.

The activity worksheets set a task in mainly mathematical terms. The worksheets can easily be adjusted to meet the specific teaching needs – the tutor can simply alter the worksheet before it is copied so that additional help can be included, or indeed, information can be deleted to make the activity more difficult. Moreover, the help sheets can be offered to students who require more guidance with the task and computer algebra system.

Help Sheets

The optional help sheets are provided to give extra assistance and clues both in terms of the mathematics and the way in which a computer algebra system can be employed. These are also photocopiable. Normally each help sheet would be distributed along with the activity sheet, but for students already familiar with computer algebra and at the stage where they are developing confidence in its use, it would not be necessary. The separation of the two sheets enables the teacher or lecturer to decide what is best for their particular students.

Teaching Notes

The (non-photocopiable) teaching notes for each activity provide some of the following information, according to the nature of the activity:

- background notes on the activity
- solutions to the problems and tasks set
- discussion points arising from the mathematics
- indication of potential student difficulties
- suggestions for extension activities

Using computer algebra systems (CAS)

Computer algebra systems (CAS) are based on software which carries out algebraic manipulation. The software also carries out the rules of calculus and extends to all areas of algebraic mathematics, such as vectors and matrices. A CAS works algebraically. It even treats numbers 'algebraically' i.e. it works with exact values such as 1/3 and $\sqrt{2}$, rather than decimal approximations, though it can supply answers in approximated decimal form if required.

As well as handling algebra and symbolic manipulation, a CAS will plot accurate graphs of a range of functions. The ability to link the algebraic work with graphs provides a powerful method of helping students in their mathematical thinking and problem solving.

There are a number of different CAS available – all differing in their user interface. However, they all have the same types of mathematical functionality. They differ mainly in the way the user types in instructions or uses the mouse – and they differ in some of the additional facilities they make available.

The appendices describe in detail how different systems are used to carry out these specific common commands. The packages we include are:

- Derive
- MathPlus/Theorist
- Maple
- Texas Instruments TI-92
- Mathematica
- Macsyma

We have identified 12 common command processes:

- **APPROXIMATE** – generates the decimal approximation of an exact number.
- **EXPAND** – multiplies out brackets and provides an expanded expression.
- **FACTORIZE** – the reverse of 'expand', groups an algebraic expression into a product of (possibly bracketed) expressions.
- **DIFFERENTIATE** – determines the algebraic derivative of a function – provided the derivative exists.
- **INTEGRATE** – produces the integral of a function. For indefinite integrals, an algebraic solution will be produced if an appropriate anti-derivative exists. Otherwise, most CAS generate definite integrals by numerical methods.
- **SOLVE** – produces the solution of an equation or set of equations. If the equation(s) cannot be solved by an exact method, then most CAS can carry out a numerical method of solution.
- **PLOT** – generates the graph of a given function.
- **SCALE/ZOOM** – for adjusting the x and y scales of a graphed function. Some CAS's will automatically scale the y-axis for a given domain.
- **LIMIT** – generates the limits of sequences or functions, given a limit such as a real number or $\pm\infty$.
- **SIMPLIFY** – rearranges an expression into a 'simpler' form. On some CAS's, this is used to generate the final results of operations such as 'differentiate'.

- **SUM** – generates a closed form expression for the sum of a (finite or infinite) series. Where no such expression exists, it will generate the numerical result for a definite sum.
- **SUBSTITUTE** – substitutes either a number, variable or expression for a given variable in a target expression.

All the worksheets in this book use these 12 specific commands, and the appendices provide a summary of how these commands are implemented on some common computer algebra systems available when we produced this book. We have also included the keyword **INPUT** in the activities, which is straightforward in all the CAS we are familiar with. Naturally, all CAS have further features and short-cuts which the student will gradually learn as experience is developed.

Many students can be put off by the bewildering array of commands and facilities available when they meet a CAS for the first time. It is no good trying to teach students every single possible function of a CAS before starting work on mathematics – it would take too long! Also students who are learning mathematics will not fully understand many of the CAS options since they will not yet have learnt the underlying mathematics.

An effective approach to teaching students is to teach them how to carry out the 12 basic processes on their particular CAS. With this knowledge it is possible to start work on mathematical problems. Further CAS commands and facilities can be introduced later – indeed, once pupils have started to work with their CAS they become quite adept at learning the additional features and commands themselves. It is surprising how much can be done with a knowledge of only the basic facilities of the CAS.

The material in this book is intended to be used as a supplement to normal teaching and each worksheet can be included at a suitable point in a teaching scheme for a particular topic. The basic part of most activities will take an hour or two to complete – many have suggestions for follow-up work which can represent a more extensive investigation suitable for private study.

The organisation of CAS work with students depends very much on the way computer resources are organised. In most cases computers are only available in a separate 'lab' environment and the CAS activities have to be seen as class 'lab' work, either for individuals or groups. In private study time, individuals can pursue specific work on their own. When hand-held CAS are more widely available this style of teaching will be supplemented by a more continuous use of CAS.

Further Reading

ATM, (1993), *Computer Algebra Systems,* Micromath 9(3), p.17–42.

This edition includes a number of articles on CAS in the classroom. Micromath often publishes teaching ideas, articles and features on computer algebra for use in schools.

Berry, J., Graham, E. and Watkins, A., (1993), *Learning Mathematics through Derive,* Chartwell-Bratt.

A relatively traditional A-level pure maths textbook with Derive incorporated throughout.

Goldstein, R. (ed), (1995), Algebra at A-level – *How the Curriculum Might Change with Computer Algebra,* Association of Teachers of Mathematics (ATM)

This volume brings together the ideas of a group of experienced A-level mathematics teachers. It addresses itself to the medium and long-term future of CAS and covers teaching, curriculum and assessment issues.

The International Derive Journal

Though focused on Derive, there is a substantial emphasis on teaching issues which are of general relevance. Editor: J. Berry, Centre for Teaching Mathematics, University of Plymouth, Plymouth PL4 8AA.

Karian, Z. (ed), (1992) *Symbolic Computation in Undergraduate Mathematics Education,* Mathematical Association of America (MAA)

A collection of articles on integrating CAS into US undergraduate teaching, including case-studies on calculus, differential equations and statistics.

Leinbach, C., (1990), *Calculus Laboratories using Derive,* Wadsworth.

At the level of A-level and slightly beyond. 20 activities with gapped worksheets written mainly for the US market, with material for A-level and slightly beyond.

Monaghan J. and Etchells T. (eds), (1993), *Computer Algebra Systems in the Classroom,* Centre for Studies in Science and Mathematical Education, University of Leeds.

This volume includes a number of articles on how CAS was integrated into mathematics teaching in a number of UK classrooms – including at GCSE level.

Taylor. M., (1995), *Calculators and CAS and their use in mathematics examinations,* Mathematical Gazette, Vol. 79, No. 484.

A thorough and practical account of the implications of graphic calculators and CAS on A-level examinations today.

Multiplying Straight Lines

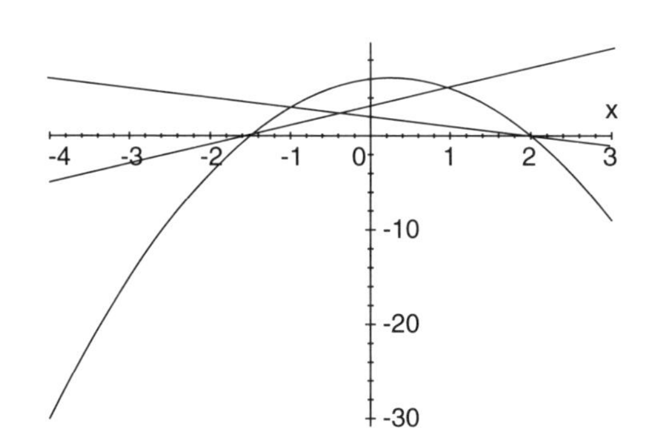

(1) Plot the graphs of the linear functions $y = x + 3$ and $y = 2 - x$. Then plot the product of these two functions; $y = (x + 3)(2 - x)$. This type of graph is called a parabola. Explain the shape of the parabola in terms of the two straight line graphs.

(2) Plot the following pairs of linear functions and – before plotting the product of each pair – predict the following features of the curve:
(a) where the curve crosses the x and y axes
(b) the co-ordinates of the turning point
(c) the 'shape' of the parabola; i.e. whether it has a maximum or a minimum

i) $(x + 3)$ & $(x + 4)$ ii) $(x - 2)$ & $(x + 3)$ iii) $(2x + 1)$ & $(x - 2)$
iv) $(5 - x)$ & $(5 - x)$ v) $(6 - 2x)$ & $(3x + 6)$ vi) $(3 - x)$ & $(2 - x)$

(3) Given the function $y = (Mx + P)(Nx + Q)$, find the following in terms of M, N, P or Q (you do not need to use all the variables in each case):
(a) the value of y when $x = 0$
(b) the value(s) of x where the function crosses the x-axis
(c) the co-ordinates of the minimum or maximum point

What values are required for the parabola to have a minimum rather than a maximum?

(4) By altering pairs of linear functions and plotting their product, can you get:
(a) the turning point of a curve directly above the intersection of the lines
(b) the curve to touch the x-axis
(c) the curve to move down one unit?

(5) Plot the quadratic function $x^2 + 2x - 3$; this also gives a parabola! Can you find two linear functions which – if multiplied together – give the same parabola? Factorise $x^2 + 2x - 3$ to check your prediction.

Multiplying Straight Lines

(3) (a) **SUBSTITUTE** $x=0$ into $(Mx + P)(Nx + Q)$

(b) If the function crosses the x-axis, the y value must be zero – which means $Mx + P = 0$ or $Nx + Q = 0$

(c) The x value of the turning point is always the midpoint between the roots of the two straight lines

(4) (a) Remember, the x-value of the maximum or minimum is the mid-point between the two roots. So choose two straight lines which intersect at this same x-value. Secondly, the y value of a maximum or minimum needs to be greater than the y value of the intersection point.

(b) If the parabola is to touch the x-axis, the y co-ordinate at the maximum or minimum needs to be zero.

(c) Explore what happens to the curve if you change the two constants of the straight line graphs.

(5) Examine the x values where the parabola crosses the x-axis – what does this tell you about each linear factor? Examine the y values where the parabola crosses the y-axis. What does this tell you about the y values of each linear factor at $x = 0$?

Teaching notes

Background

Part of this activity was taken from an idea by Paul Drjivers. It can be used as an introduction to quadratic functions and graphs, as well as an introduction to methods of solving quadratic functions. Although the activity itself can be used with a graphing technology, it can serve as an introduction to the graphing capabilities of the CAS you use.

Solutions

(2) Graphs of the functions are given below.

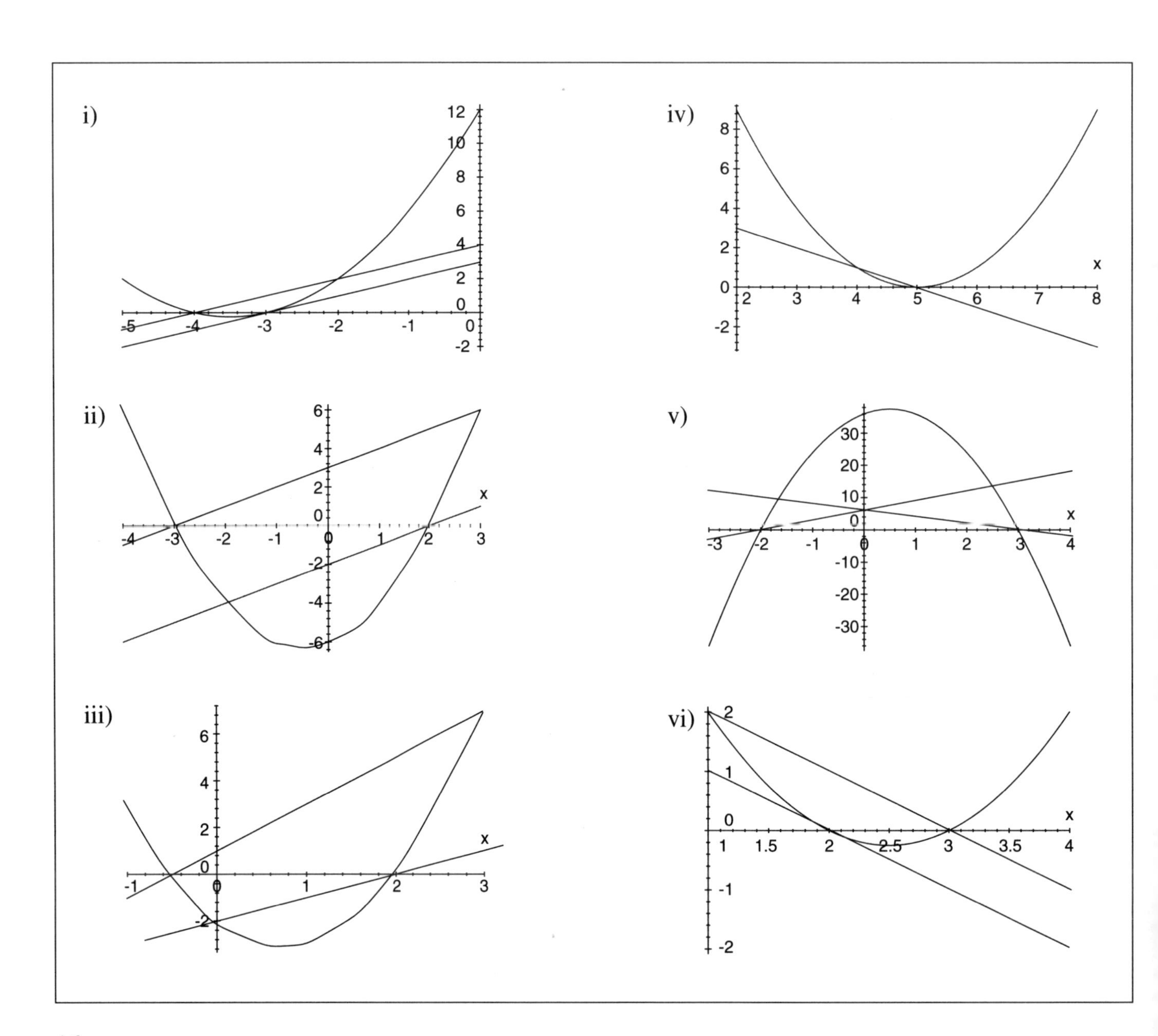

(3)
- when $x = 0$, $y = P.Q$
- when $y = 0$, $x = -\frac{P}{M}$ or $x = -\frac{Q}{N}$
- The co-ordinate of the turning point is
  $(-\frac{1}{2}(\frac{P}{M}+\frac{Q}{N}),(-\frac{M}{2}(\frac{P}{M}+\frac{Q}{N})+P)(-\frac{N}{2}(\frac{P}{M}+\frac{Q}{N})+Q))$
- The parabola has a minimum when $M.N > 0$

Misconceptions

There may be some difficulties over the notation; it should be made clear that, '*y* = …' or '*f*(*x*) = …' are both perfectly good notations for functions. There may also be difficulties relating the linear factors to whether the quadratic has a maximum or a minimum.

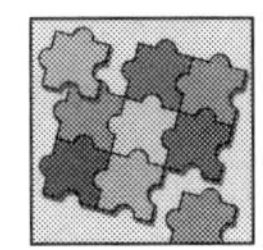

Extensions

To extend part (5), students can explore other quadratics of the form $x^2 + bx + c$ – and – if possible – find the linear factors in each case. If some quadratics have no real roots, and hence no real factors, students can formulate a way of testing whether a quadratic can be factorised or not.

Equation of a Tangent

(1) Find the equation of the tangent to the curve $y = 1 - x^2$ at the point (1,0).

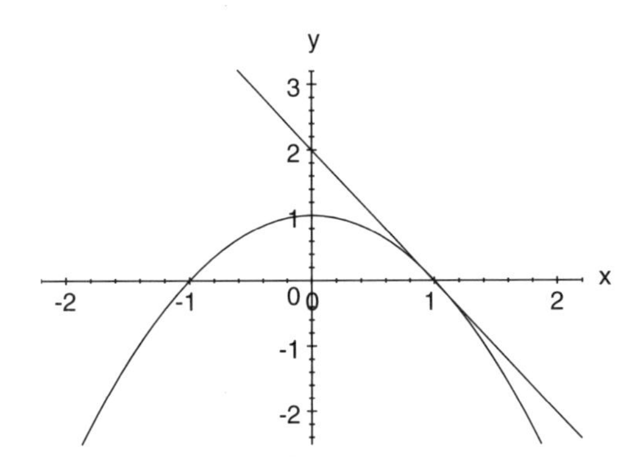

You will need to calculate the gradient of the curve m at $x = 1$ using calculus to differentiate. Then use the equation of a straight line given the gradient and a known point $(x_1,y_1,)$ on the line, $y - y_1 = m(x - x_1)$.

(2) Plot both the curve and your answer to check that it works!

(3) Plot the tangents to the curve $y = 1 - x^2$ at the points whose x co-ordinates are –3,–2,–1,0,2 and 3.

(4) Find the equation of the tangent to these curves at the given points

(a) $y = x \sin x$ at $x=1$

(b) $y = \dfrac{1}{1 + x}$ at $x=1$

(c) $y = \dfrac{4x}{\sqrt{1 + x^2}} - 5x + 2$ at $x=2$

Extension

(5) Find the equation of the tangent to the curve $y \cos x + x \sin y = 1$ at the point (0,1).

Equation of a Tangent

Help Sheet

We will work through example 4(a), i.e. plot the tangent to the function $f(x) = x\sin x$ at the point x=1.

(i) **INPUT** the expression $x\sin x$.

(ii) To find the gradient of $f(x)$ you should **DIFFERENTIATE** with respect to x, order 1.

(iii) To find the gradient of $f(x)$ at x=1 **SUBSTITUTE** x=1 into the expression for the derivative.

(iv) To construct the equation of the tangent **INPUT** $y - \sin 1 = (\cos(1) + \sin(1))\,(x - 1)$.

(v) Now **SOLVE** the equation in (4) for y to obtain $y = (x - 1)\cos(1) + x\sin(1)$, or equivalent.

(vi) Now **PLOT** the original function and its tangent. If you **APPROXIMATE** the trigonometric expression for the tangent you should have $y = 1.38176x - 0.540229$.

Teaching notes

Background

The equation of a straight line (the tangent) is $y = mx + c$, where

m is the gradient of the straight line

c is the y co-ordinate of the intersection of the straight line and the y axis.

The gradient of a tangent to a curve $y = f(x)$, at the point (x_1, y_1), can be calculated by differentiating the equation of the curve and substituting x_1 into it. The equation of the tangent can be constructed from the equation $y - y_1 = m(x - x_1)$.

Solutions

(1) $y = -2x + 2$

(3) $y = 6x + 10$, $y = 4x + 5$, $y = 2x + 2$,
$y = 1$, $y = -4x + 5$ & $y = -6x + 10$

4(a) $y = 1.38176x - 0.540299$

4(b) $y = \dfrac{(3 - x)}{4}$

4(c) Approximately $y = 4.86216 - 4.64222x$

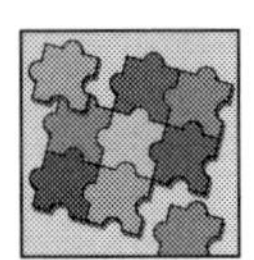

Extensions

This is a little more demanding than the previous examples. Computer algebra systems can be made to differentiate implicitly, but if your students are beginners they may not be aware of this. If they differentiate the expression as it stands, with respect to x, the CAS may treat y as constant and hence obtain an incorrect answer! Depending on how familiar your students are with (a) their CAS or (b) the technique of implicitly differentiating, you might suggest using pencil and paper for parts of this work! Differentiating implicitly we obtain

$$x\cos y\frac{dy}{dx} + \sin y + \cos x\frac{dy}{dx} - y\sin x = 0$$

Making $\frac{dy}{dx}$ the subject we obtain $\dfrac{dy}{dx} = \dfrac{y\sin x - \sin y}{x\cos y + \cos x}$

substituting x=0 and y=1 we have

$$\frac{dy}{dx} = -\sin(1)$$

which gives the tangent as $y = -\sin(1)x + 1$

Most CASs can plot implicit functions, the figure below shows a plot of the implicit function and is tangent at x=0.

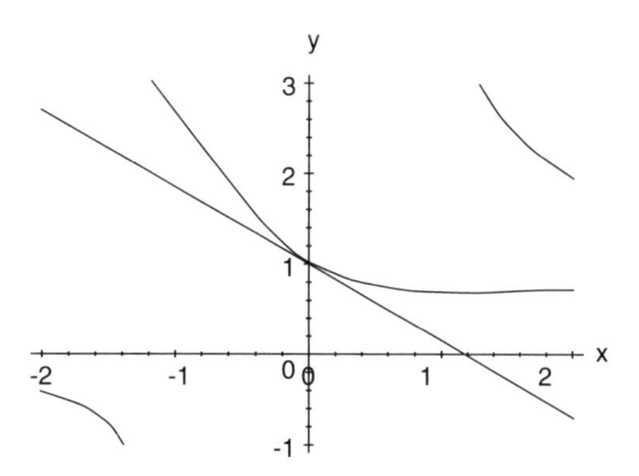

Taxing Functions

Taxation on earnings has always been treated as a progressive taxation i.e. the more you earn the higher proportion of tax you pay. For example, national insurance and income tax – the two taxes on earnings in the U.K. – are calculated differently:

National Insurance: Monthly earnings are split into two bands:
income up to £144 is taxed at 2%.
income between £144 and £270 is taxed at 10%.
income above £270 is not taxed

Income Tax: Everyone has a tax allowance – £3140 per annum for most people. Any earnings beyond this allowance is your taxable income, and is taxed by splitting it into three tax brackets:
taxable income up to £2000 is taxed at 20%
taxable income between £2000 and £18500 is taxed at 25%
taxable income above £18500 is taxed at 40%

For example, if you earn £1100 per month; i.e. 12 x £1100 = £13200 per year:

National Insurance		Income Tax	
2% of £144	= £2.88	Taxable Income	= £13200 – £3140 = £10060
10% of £126	= £12.60	20% of £2000	= £400
Monthly Total	= £15.48	25% of £8060	= £2015
Yearly Total	= £15.48 x 12 = £185.76	Yearly Total	= £2415 per year

(1) Define two functions – ni(x) and it(x) – which take annual income as input and calculate national insurance and income tax respectively. Test these functions carefully with different inputs.

(2) Plot the functions between 0 and 30000. How do the shapes of the two functions compare? Which of the two functions gives the fairest tax system? Explain your answer.

(3) Use the ni(x) function to define a further function – nirate(x) – which takes annual income as input and calculates national insurance as a percentage of annual income. Define an equivalent function – itrate(x) – which takes annual income as input and calculates income tax as a percentage of annual income.

(4) By examining the graph of nirate(x), find the annual income of someone who pays the maximum percentage of national insurance. What does this tell you about the national insurance tax system?

(5) Examine the graph of itrate(x). What is the maximum tax rate? How does the income tax system compare with the national insurance system?

Taxing Functions

Help Sheet

(1) Below is an example of an imaginary tax function which is defined over different domains.

$f(x) = 0.1x$	if $x \leq 110$	10% on earnings upto £110
$= 11 + 0.2(x - 110)$	if $x \leq 200$	Plus 20% on earnings between £110 and £200
$= 29 + 0.3(x - 200)$	otherwise	Plus 30% on earnings above £200

On some systems, a function like this can be defined using nested IF statements. For example, on Derive:

$f(x) := \text{if}(x \leq 110, 0.1x, \text{if}(x \leq 200, 11{+}0.2(x{-}110), 29{+}0.3(x{-}200)))$

Note: the definition of national insurance is given on a monthly basis on the activity sheet, while the ni(x) function should output the *annual* national insurance.

(2) **PLOT** the functions ni(x) and it(x) separately, because one function has a greater range than the other. For both functions, start with a domain of $£0 \leq x \leq £30000$ – which covers the majority of people's incomes. For the ni(x) function, **PLOT** the function with a range of $0 \leq y \leq 200$ and use a range of $0 \leq y \leq 15000$ for the it(x) function. A graph of a progressive tax system should increase as the income increases. A regressive tax system either increases at the same rate or flattens out so that beyond a particular income, the tax is a fixed amount.

(3) The nirate(x) and itrate(x) functions should give values between 0% and 100%.

(4) Important features to examine are how the rate changes as the income increases and whether there are any maximum rates of tax. This last feature would mean the tax rate would peak at a particular income, then decrease for higher incomes – a typical feature of a regressive tax system.

(5) Is the maximum tax rate for income tax ever reached?

Teaching notes

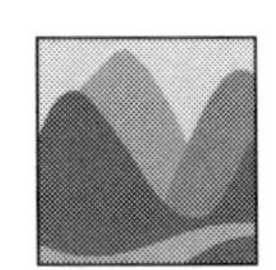

Background

Taxation functions are interesting because they are 'real' piece-wise defined functions. If you are using this activity outside the U.K, you may want to adapt the worksheet and use the taxation system in your own country as the basis for the functions; for example, in the U.S.A, the two functions could be Federal and State income tax.

Solutions

(1) There are several ways of defining the ni(x) function. For example, an intermediate function can be defined for monthly incomes:

$$
\text{monthni}(x) = \begin{cases} 0.02\,x & \text{if } 0 \le x \le 144 \\ 2.88 + 0.1(x-144) & \text{if } 144 < x \le 270 \\ 15.48 & \text{otherwise} \end{cases}
$$

The yearly national insurance can then be calculated with the following definition:

$$\text{ni}(x) = 12\ \text{monthni}(x/12)$$

The income tax function can be defined as follows:

$$
\text{it}(x) = \begin{cases} 0 & \text{if } 0 \le x \le 3140 \\ 0.2(x-3140) & \text{if } 3140 < x \le 5140 \\ 400 + 0.25(x-5140) & \text{if } 5140 < x \le 23640 \\ 5025 + 0.4(x-23640) & \text{otherwise} \end{cases}
$$

(2) The ni(x) function, clearly indicating the two *rates* of tax and a constant tax after an annual income of around £3200:

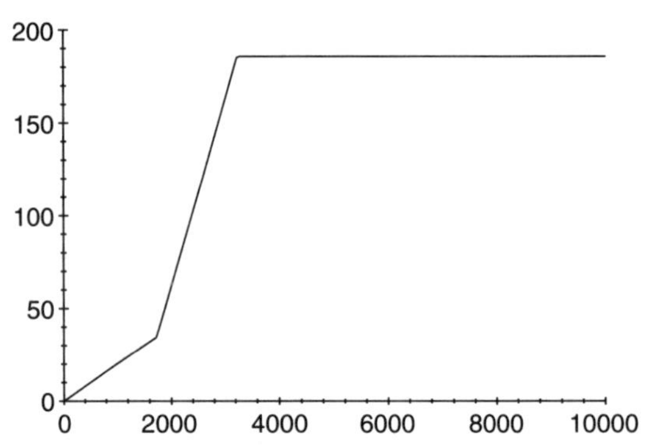

The it(x) function, flat up to the tax allowance, then shows a continuous function with three fixed gradients, representing the three tax bands:

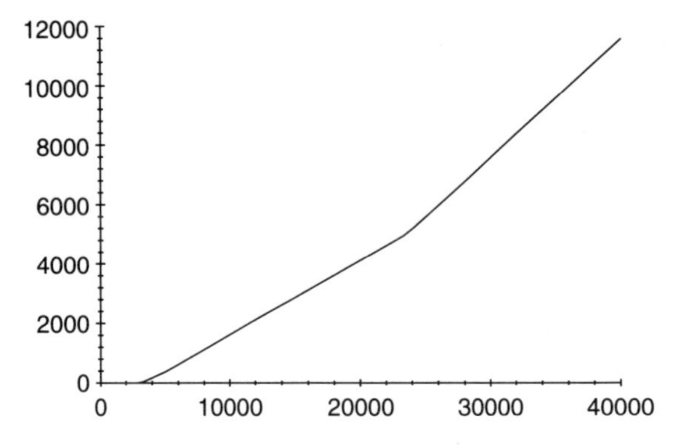

(3) The functions nirate(x) and itrate(x) – given as percentages – are defined as:

nirate(x)	$= 100\ \text{ni}(x) / x$
itrate(x)	$= 100\ \text{it}(x) / x$

The graph of nirate(x) is fixed at 2%, then increases to a maximum, while the function tends to 0 as annual income increases. The taxation is therefore regressive.

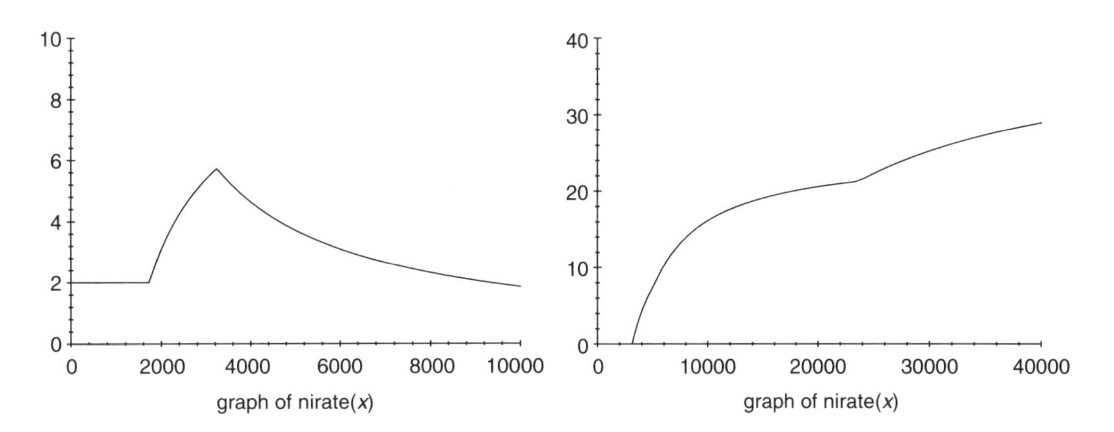

The itrate(x) function indicates a progressive taxation system; though the rate increases rapidly for small incomes beyond the tax allowance. The graph has an assymptote at y = 40, which represents the maximum tax rate – although it is never reached!

Misconceptions

There may be a number of problems with definitions of piece-wise functions. Some students may not believe they are real functions, as they have been used to functions defined by single formulas. This misconception would provide a good opportunity to revise the mapping definition of functions. Students may also have difficulty with the CAS syntax for piece-wise functions, so some preliminary work on defining more familiar piece-wise functions (e.g. $y = |x|$) may be useful. There may be some confusion between monthly and annual income, and what is required to convert the monthly defined national insurance into an annually defined function.

Discussion

There is plenty of scope for discussion in this activity – especially around the issue of what a 'fair' taxation system is – and what the graph of a fair tax function should look like. Students – especially those studying Economics or Business Studies – may want to go and read about taxation systems – in order to make better sense of their findings; e.g. what is the formal definition of a 'progressive' vs. a 'regressive' taxation system. Other questions could be; why is national insurance set up the way it is? What would be the consequences of removing the tax allowance?

A manufacturer is designing square floor tiles with two curves making up a diagonal leaf shape – with the leaf in green and the surround in red. One possible design is where the two curves trisect the angles at the corners (in other words, all the corner angles are 30°). This is so that the leaf patterns continue smoothly from tile to tile.

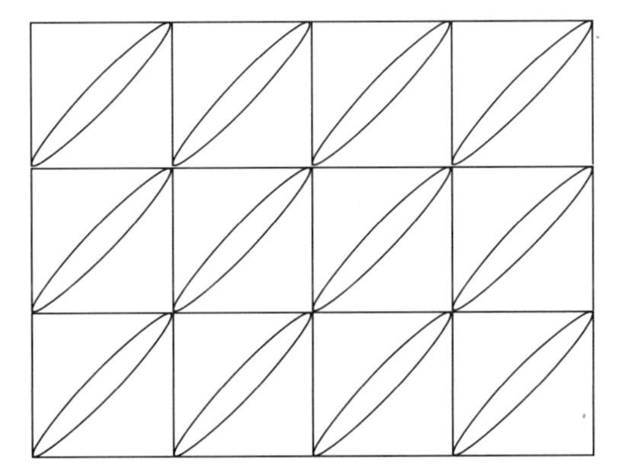

(1) If the two curves are defined by cubic functions, determine the formulas of each. You can assume that the tile covers the unit square, so having the vertices at (0,0), (1,0), (1,1) and (0,1). Plot your solution curves to make sure the solutions 'look' right.

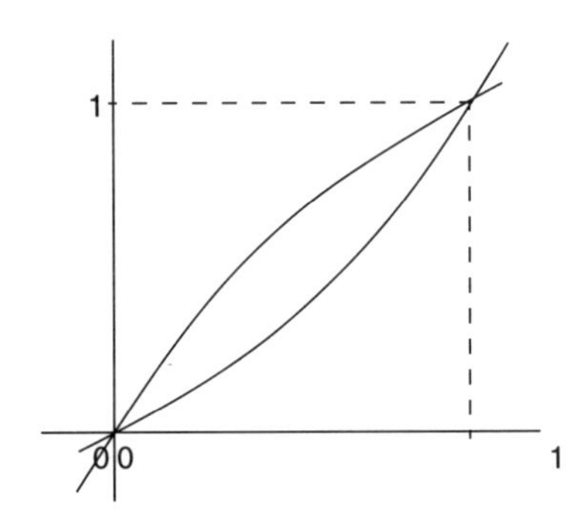

(2) Calculate the relative amounts of the two colours in the design.

(3) Another possible design is to add the constraint that the three areas of the tile have to be equal. Determine two polynomial functions for the curves in this case. Examine the curves and assess whether they make a good design.

(1) Starting with either curve, one method of solution is to **INPUT** a 'template' solution; i.e. a cubic of the form:

$ax^3 + bx^2 + cx + d$

The idea is to determine values for *a, b, c* and *d* from what you know about the curve. Given you have 4 unknowns, you are going to need four equations with these unknowns, which you can then **SOLVE** simultaneously to give you possible values for *a, b, c,* and *d.*

From the tile pattern, it should be possible to 'extract' four features of each curve which can be used to construct four equations. For example, the polynomial has the value 1 when $x = 1$, so you can **SUBSTITUTE** this value into the expression and generate the equation:

$a1^3 + b1^2 + c1 + d = 1 \Rightarrow a + b + c + d = 1$

Some of the features involve the gradients of the curves; for example, the lower curve has the gradient $\tan(\pi/6)$ at 0. Finally, once you use **SOLVE** on the four equations to determine *a, b, c* and *d,* substitute the values back into the original equation and **PLOT** this to see whether it 'looks' right.

(2) To determine the relative areas on the tile, you will need to **INTEGRATE** your two solution polynomials between the appropriate limits and determine the area between the two graphs.

(3) The method by which this problem can be solved is essentially the same as in (1). For example, if you have a polynomial for the lower curve, then this still has to go through the points (0,0) and (1,1) and have appropriate derivatives at $x = 0$ and $x = 1$. The area constraint adds a fifth equation; i.e. the definite integral – between the limits 0 and 1 – has to come to a particular value – which value?

Teaching notes

Background

This activity was taken from an idea in 'Symbolic Computation in Undergraduate Mathematics Education' edited by Zaren Karian. The activity is based on modelling curves to fit a number of constraints and using CAS equation solving capabilities to determine the coefficients of polynomials from a set of equations.

Solutions

(1) If the general formula is $f(x){:=}ax^3{+}bx^2{+}cx{+}d$, then the four equations required for the lower curve are:

$f(0)=0$	so	$d = 0$
$f(1)=1$	so	$a + b + c + d = 1$
$f'(0)=\tan(\pi/6)$	so	$c = \tan(\pi/6)$
$f'(1)=\tan(\pi/3)$	so	$3a + 2b + c + d = \tan(\pi/3)$

Similarly for the upper curve;

$f(0)=0$	so	$d = 0$
$f(1)=1$	so	$a + b + c + d = 1$
$f'(0)=\tan(\pi/3)$	so	$c = \tan(\pi/3)$
$f'(1)=\tan(\pi/6)$	so	$3a + 2b + c + d = \tan(\pi/6)$

Solving each of these solution sets and substituting into the original formulas should give the following two polynomials:

$$\left(\frac{4\sqrt{3}}{3} - 2\right)x^3 + \left(3 - \frac{5\sqrt{3}}{3}\right)x^2 + \frac{\sqrt{3}}{3}x$$

$$\left(\frac{4\sqrt{3}}{3} - 2\right)x^3 + \left(3 - \frac{7\sqrt{3}}{3}\right)x^2 + \sqrt{3}x$$

(3) Integrating each of the cubic solutions above, between the limits 0 and 1, should give you the values:

$$\frac{1}{2} - \frac{\sqrt{3}}{18} \qquad \frac{1}{2} + \frac{\sqrt{3}}{18}$$

So the area of the leaf is $\frac{\sqrt{3}}{9}$ and it takes up 19.25% of the unit square.

(4) Adding the area constraint requires quartics to solve the five simultaneous equations for each function. The fifth equation should involve a definite integral of the function in question – between the limits 0 and 1. This definite integral should equal 1/3 or 2/3 – depending on whether the function represents the upper or lower curve.

Given a quartic of the form:

$ax^4 + bx^3 + cx^2 + dx + e$ then the solution for each curve is

upper curve	lower curve
$a = 5 - \frac{5\sqrt{3}}{3}$	$a = \frac{5\sqrt{3}}{3} - 5$
$b = \frac{14\sqrt{3}}{3} - 12$	$b = 8 - 2\sqrt{3}$
$c = -4\sqrt{3}$	$c = -2$
$d = \sqrt{3}$	$d = \frac{\sqrt{3}}{3}$
$e = 0$	$e = 0$

Misconceptions

The activity is quite difficult and students may encounter a number of very different difficulties. Students would benefit greatly from some initial exposition on the concepts and techniques involved. Another suggestion is that students are given some simple problems to solve for themselves, to develop their confidence with the technique on a CAS. e.g. finding quadratics which go through three particular points, finding quadratics which go through two particular points, and with a particular derivative at one point.

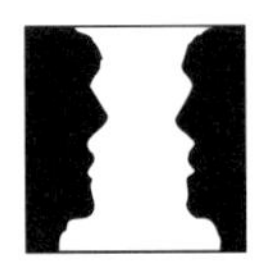

Discussion

There are a number of issues that may arise from this activity. One important issue is the order of the polynomials required to solve the problem – which is related to ideas behind the solution of sets of linear simultaneous equations. You may want to suggest to students to attempt (1) with quadratics, to see how far they get. In (4), students are left to find out for themselves whether a cubic is sufficient – or whether a quartic is required. This in itself will hopefully result in spontaneous pupil-pupil discussion.

Another issue which may arise is the idea that – given the graphs 'look' symmetrical about $y = x$, the inverse function of one of the solution curves can be used to find the other solution curve. This can be usually found using a CAS by solving $y = f(x)$ for y. This does indeed generate an appropriate curve, but with a more complex function than a cubic.

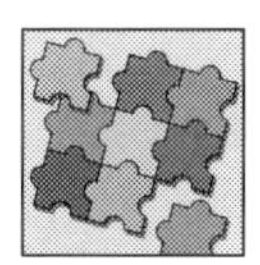

Extensions

An obvious extension is to try to model the curves with very different functions; e.g. a circle function or a trigonometic function. This is mathematically more demanding and also requires a more sophisticated use of a CAS to solve the appropriate equations.

A further extension is for students to develop their own simple curves on a tile design. To constrain the problem, students could make sure curves link together smoothly from tile to tile; i.e. that the curves continue from tile to tile and that the derivatives are the same at these points, e.g.

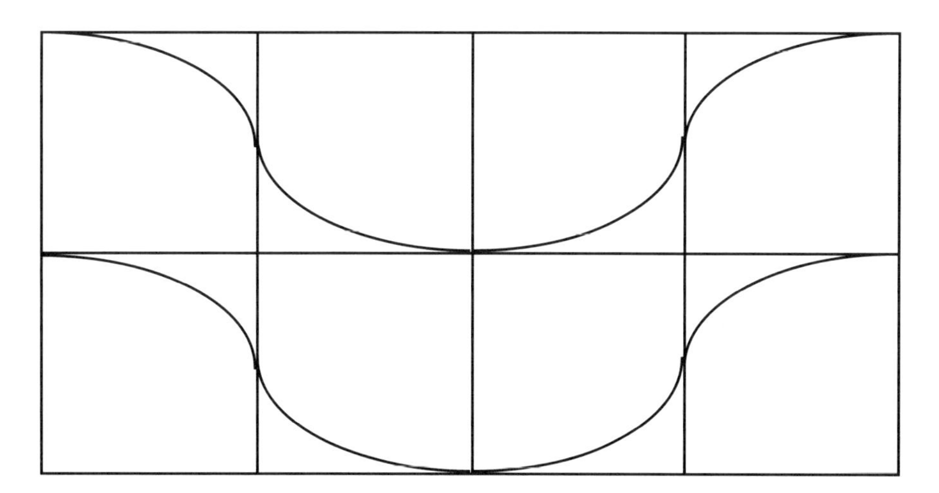

Visualising Functions and Derivatives

This activity is designed to help you visualise the shape of the graph of the derived function, $f'(x)$, and relate it to the shape of the graph of the original function, $f(x)$.

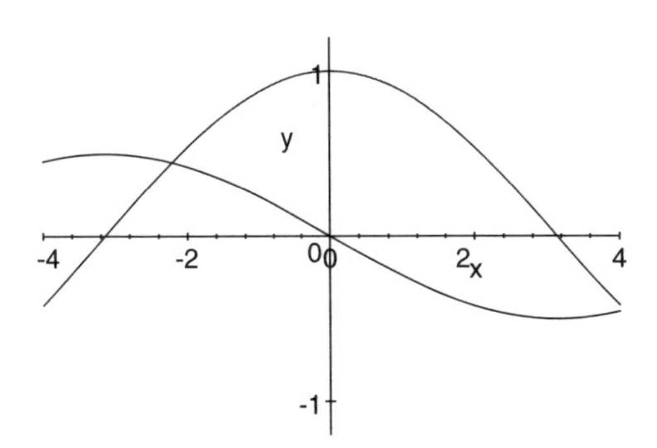

(1) Write down the equation of a function with which you are familiar, $f(x) = x^2 + 1$ e.g.. Sketch the graph of this function with roughly similar scales on both axes. Write down the equation of the derivative, e.g. $f'(x) = 2x$.

Sketch the graph of the derivative on the same set of axes.

Examine both graphs very carefully and focus on the value of the derivative and the gradient of the original function at specific x values. They should be the same, e.g. $f'(1) = 2$ and the gradient of $f'(x)$ at $x = 1$ is 2.

In the next part, focus on the value of $f'(x)$ and the gradient of $f(x)$.

(2) Answer these three questions and summarise your answers in this table

- What is happening to $f'(x)$ when $f(x)$ has a peak $\cap$ or a trough $\cup$?
- What is happening to $f(x)$ when $f'(x)$ is negative?
- What is happening to $f(x)$ when $f'(x)$ is positive?

$f'(x)$	$f(x)$
	peak or trough
negative	
positive	

(3) Repeat this activity for many different functions.

Several are suggested below but you should make some up on your own.

Are you able to predict the shape of $f'(x)$ when you know the shape of $f(x)$?

$f(x) = (x-1)(x+1)$ $\quad$ $f(x) = x(x-1)(x+1)$ $\quad$ $f(x) = x(x-2)(x+1)(x+3)$

$f(x) = \dfrac{1}{x}$ $\quad$ $f(x) = \dfrac{1}{x^2}$ $\quad$ $f(x) = \dfrac{1}{1+x^2}$

Visualising Functions and Derivatives

Help Sheet

You should try this activity after you have learnt the basic ideas of differentiation. You should know things like if $y = x^2 + 3x - 4$ then $\frac{dy}{dx} = 2x + 3$.

This activity is designed to help you visualise the graph of the derivative, $f'(x)$ and relate it to the shape of the original function, $f(x)$.

Don't let notation be a problem. $y = x^2 + 3x - 4$, $\frac{dy}{dx} = 2x + 3$ is just another way of writing $f(x) = x^2 + 3x - 4$, $f'(x) = 2x + 3$. Your computer algebra system may omit the symbols $y, \frac{dy}{dx}, f(x), f'(x)$, and simply write $x^2 + 3x - 4$ and $2x + 3$.

The purpose of this activity is for you to appreciate the visual relationship between a function and its derivative. Do not feel that it is cheating to let the computer algebra system work out the derivative for you. Whilst it is useful to be able to **EXPAND** $\frac{1}{10}(x-1)(x-2)(x+3)(x-4)$ to get $\frac{x^4}{10} - \frac{3x^2}{2} - x + \frac{12}{5}$ and then **DIFFERENTIATE** to get $\frac{2x^3}{5} - 3x - 1$, you can use your computer algebra system to do the calculations for you here so you can concentrate on the visual effects.

Your graphs may not appear. In this case the most likely problem is that they are outside your viewing area. Try **ZOOMING** out along both the x and y-axes.

If the problems of graphs being off the screen persist, then try using functions of the form

$$f(x) = \frac{1}{10}(x-a)(x-b)(x-c)(x-d)$$

where a, b, c and d are small positive or negative numbers, e.g.

$$f(x) = \frac{1}{10}(x-1)(x+2)(x+3)(x+4).$$

Teaching notes

Background

The aim of this activity is purely to assist students in gaining a visual 'feel' for the relationship between the graph of a function and the graph of its derivative. Algebraic aspects of differentiation are not addressed and the HELP SHEET even suggests that a computer algebra system may be used to assist with differentiation. This activity is designed for use early in a calculus course and the examples used in the sheet reflect this intention.

Specific objectives depend on the student's competence: an objective for a high attaining student is for him/her to be able to sketch the graph of the derivative when presented with the graph of a function; an objective for a mid-range student is for her/him to know that $f'(x) = 0$ at a maxima or minima and that $f'(x) < 0 \Rightarrow f(x)$ is decreasing and $f'(x) > 0 \Rightarrow f(x)$ is increasing.

This activity is more open than many of the activities in this book. For this reason solutions are not given, the HELP SHEET is of limited help and the need for teacher supervision is increased.

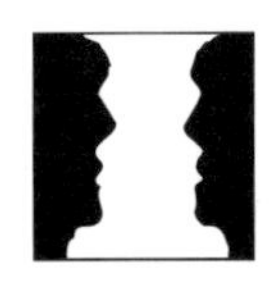

Discussion

Many teachers will not like $f(x)$ and $f'(x)$ sketched on the same axes. This can be avoided by opening two graphics windows, one for $f(x)$ and one for $f'(x)$. This is not always satisfactory as aspects of the curve are sometimes lost or one axis is compressed, especially with hand-held technology.

Misconceptions

The most likely difficulty is that the student will not focus on the intended relationship. This is an aspect of the open nature of the activity. An initial teacher-led example may be desirable.

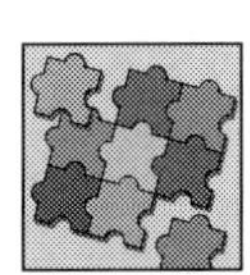

Extensions

- From the graph of a function, sketch the graph of its derivative.
- By relating $f''(x)$ to $f'(x)$ in a similar way to which the activity related $f'(x)$ to $f(x)$, examine the visual relationship between $f''(x)$ and $f(x)$.

- Examine the effect on $f''(x)$ and $f'(x)$ of functions which have the following shapes a)  b)  c)  d)

  Functions to examine include:

  $f(x) = x^3$, $f(x) = -x^3$, $f(x) = \dfrac{x^5 - x^4 + x^3 + x^2 - x + 1}{x^3 - 2x}$

  (especially around the interval (-1,0))

Activity Worksheet

The Approximate Derivative Function

The ADF gives an approximation of the steepness of the graph of a function for any value of x. It is a function which gives an approximation to the gradient of the graph at each point.

The formula for calculating an approximate derivative function is:

$$\frac{f(x+h) - f(x)}{h}$$

where h is a very small constant.

The ADF of the function $\sin(x)$ is $\frac{\sin(x+h) - \sin(x)}{h}$. Such a function will tell you the approximate steepness of the sine function at each value of x. Note that you cannot substitute $h = 0$ (Try it and see!)

(1) Plot the expression $\frac{\sin(x+h) - \sin(x)}{h}$ from $x = -2$ to $x = 2$ but first substitute $h = 0.9$.

(2) Now form a *family* of graphs by plotting $\frac{\sin(x+h) - \sin(x)}{h}$ from $x = -2$ to $x = 2$ but substitute in turn, $h = 0.9, 0.8, 0.7, \ldots, 0.2, 0.1$. Show all these graphs together on the same axes. What do you see as you substitute smaller and smaller values of h?

(3) Write a brief report to describe your results.

(4) Repeat this work for the ADF of the function $-\frac{x^3}{15} + \frac{x^2}{6} + \frac{11x}{10}$, namely

$$\frac{\left[-\frac{(x+h)^3}{15} + \frac{(x+h)^2}{6} + \frac{11(x+h)}{10}\right] - \left[-\frac{x^3}{15} + \frac{x^2}{6} + \frac{11x}{10}\right]}{h}$$

(5) Repeat again for a function of your own choosing.

The Approximate Derivative Function

Help Sheet

The formula for calculating an approximate derivative function is:

$$\frac{f(x+h) - f(x)}{h}$$

(1) **INPUT** the expression $\frac{\sin(x+h) - \sin(x)}{h}$ and **SUBSTITUTE** $h =$ 0.9. Then **PLOT** the graph from $x = -2$ to $x = +2$.

(2) You can **SUBSTITUTE** the other values of h and then superimpose the **PLOT**s. On some computer algebra systems there are short cuts to creating a family of graphs. For instance, on derive you can use the vector command:

VECTOR((sin ($x + h$) – sin (x))/h, h, 0.1, 0.9, 0.1)

(3) In your report, explain what you have done and the results you have obtained. You might also include a graph of $f(x) = \sin x$ so you can explain how the ADF describes the steepness of $\sin x$ at different points. You might also plot the graph of $\cos x$ since you will see that the limit of the different ADFs has the same graph.

(4) Make it easy to create this large expression and first INPUT the cubic expression $-\frac{x^3}{15} + \frac{x^2}{6} + \frac{11x}{6}$. Then create a second expression by substituting $(x+h)$ in place of x. Finally build up the required expression from the previous two.

Teaching notes

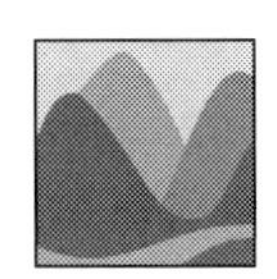

Background

This activity provides useful material for understanding the idea of the gradient of a smooth curve. Some initial discussion and exposition would help clarify what is being calculated.

The purpose of this activity is to develop students' understanding of the definition of derivative. Often this definition is related to the steepness of the graph at a particular point and the entire discussion focusses on just one point. The approach here, in contrast, focusses on the graph of the function, and makes it clear that the derivative is a *function* derived from the original function and not just a single steepness.

This approach is made possible by the fact that a computer algebra system permits easy interchange between algebraic and graphical work.

Solutions

The graphs will show a series of curves which converge to a single shape as h becomes smaller and smaller. In the report, students should sketch these curves and mention that the limit of the graphs will be the graph of the derivative.

Plotting the graph of $\cos x$ will illustrate that the derivative of $\sin x$ is $\cos x$

Question (4) again shows that a limiting shape is obtained as the values of h become smaller and smaller.

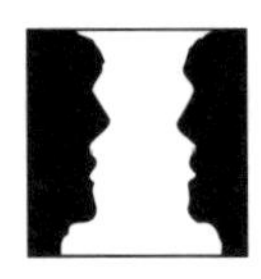

Discussion

Suggest that students simplify the expression given in (4) to obtain the expression $-\dfrac{6x^2 + 2x(3h-5) + 2h^2 - 5h - 33}{30}$. They can easily see for themselves what expression would be the limit as h approaches zero – they can substitute $h = 0$ to find the limit. So they will have worked out the derivative from first principles.

However, in (2) the expression $\frac{\sin(x+h) - \sin(x)}{h}$ does not simplify if $h = 0$, some work on expanding the trigonometric functions and taking limits is required to fully understand that the limit is cos x. If you wish to have a follow-up discussion you may prefer to suggest functions in part (5) rather than leave free choice to the student. In this case, modify the worksheet before photocopying.

(1) Use your computer algebra system to help you sketch graphs of the functions given below. As well as the graph plotting facility, you can use calculus to help you identify maxima, minima and points of inflexion.

- Remember, you must produce a *sketch*, on which you mark the key features of the graph of the function.

$$f(x) = x^3 e^{-2x}$$

$$p(x) = x^3 + 8x^2 + x$$

$$h(x) = (x^2)^{(1/3)}$$

$$g(x) = \frac{(x-3)}{(x-2)(x+1)}$$

$$q(x) = \begin{cases} \dfrac{e^{-x^2}\sin x}{x} & \text{for } x \neq 0 \\ 1 & \text{for } x = 0 \end{cases}$$

$$z(x) = \begin{cases} x^2 \sin\left(\dfrac{1}{x}\right) & \text{for } x \neq 0 \\ 0 & \text{for } x = 0 \end{cases}$$

(2) Discuss the statement:

> When you solve the equation $f'(x) = 0$ the values of x will tell you where the function $f(x)$ has local maxima and minima and points of inflexion.

This is not entirely true, as your work in part (1) will show.

Sketching Graphs

Help Sheet

(1) To help you sketch a graph, use your computer algebra (if you need it) to help you with the following.

- **PLOT** a graph of the function. Remember that it is sometimes impossible to set a scale which will show all the features of a graph on one picture. Your sketch need not be exactly to scale and so it can be made to show everything clearly.
- Use **DIFFERENTIATE** and **SOLVE** to find places where the function has zero derivatives. This will guide you in finding local maxima and minima.
- Use **DIFFERENTIATE** and **SOLVE** to find places where the **second derivative** is zero. This will guide you in finding points of inflexion.

(2) In discussing the statement, use examples from your sketches to illustrate points you make. For example, zero derivatives will often correspond to places with local maxima, minima and points of inflexion. However, there might be a local minimum where there is no derivative – or a point of inflexion where the derivative is not zero. Write a 'statement' of your own which replaces the one given.

Teaching notes

Background

This activity serves two purposes. The first is to provide some useful practice in sketching graphs making full use of computer algebra to support the sketches by the use of calculus.

Very often, the accurate graph drawn by the CAS cannot show all the features of a graph because some are very small. For instance the function $q(x) = \dfrac{e^{-x^2}\sin x}{x}$ has oscillations caused by the sine function, but a standard plot of the function on a CAS will not show them as they are too small. If you zoom in to see the oscillations then you lose the main part of the graph. You can do a sketch yourself which is not strictly to scale – but it can be made to show all the features!

Potential Student Difficulties

f(x) is reasonably straightforward with opportunity to use the second derivative to find a point of inflexion. Note that one point of inflexion occurs where the first derivative is zero but the other two points of inflexion do not. Therefore it is **not** true to imply that the first derivative will tell you where to find points of inflexion.

One of the features of *p(x)* is close to the origin and requires **ZOOM** to see it clearly. The use of calculus is helpful in finding out its exact co-ordinates. A sketch drawn 'not to scale' is necessary to show the graph's features in full.

The curious feature of *h(x)* is that it has a 'cusp'. There is no derivative at the origin – it is similar to the function $|x|$ in this respect. However there is clearly a local minimum! It therefore shows that the derivative being zero is not an entirely reliable way to find local maxima and minima. Here we have a local minimum, but no derivative and so putting the derivative equal to zero will not find this local minimum.

The function *g(x)* is relatively straightforward but has asymptotes at 2 and –1. It illustrates that the local maxima and minima are not in fact the (global) maxima and minima. Indeed there is no largest or smallest value even though lots of values are larger and smaller than the local maxima and minima!

The function *z(x)* is most peculiar and usually reserved for inclusion in more advanced mathematics books! Note that the formula must not be used for $x = 0$. Its oscillations increase in frequency as you get closer and closer to the origin. Consequently it defies most CAS in producing a fully accurate graph. As you zoom into the origin you see more and more oscillations. If you work with the definition of derivative you can prove that at the origin the derivative is indeed zero! So here is a place where there is **no** local maximum or minimum and yet the derivative is zero.

The function $q(x)$ looks 'bell-shaped' but in fact has small oscillations for larger values of x. Putting the derivative equal to zero and solving will produce equations which can only be solved numerically – so a graphical approach is important. What is really important is to realise that the sine function has zeroes and use this fact to help with the sketch.

Discussion

The statement given in (2) is not entirely true.

If there is a local maximum or minimum, and if a derivative exists, it **is** true that the derivative $f'(x) = 0$. But there can be a point of inflexion where the derivative is not zero – for example in function f. The criterion for a point of inflexion is that the second derivative is zero. Of course, some points of inflexion do occur where the derivative is zero.

Function h illustrates that you can have a local minimum, but the derivative is not zero since there is no derivative at that point – it is a 'cusp'.

The analysis of function z is rather advanced but at $x = 0$ the derivative is actually zero. This can be shown using the definition of derivative and noting that $z(0) = 0$. Though there is a derivative, and it is zero, there is no maximum, minimum nor point of inflexion at $x = 0$.

People use the condition that the derivative is zero as the first step in finding local maxima and minima. This is still a useful strategy but it is important to note that there is not such a clear correspondence as it might at first seem.

The only generally accurate statement to make is "If there is a local maximum or minimum, and if a derivative exists, then the derivative is zero."

Population and Pollution

A model for the anticipated population growth of a small town development is as follows (p is the number of people after t years)

$$p(t) = \frac{50000e^{0.6t}}{9 + e^{0.6t}}$$

A conservation consultant uses a pollution index to help predict the effect of urbanisation on the environment. The consultant uses a simple model to estimate the value of the pollution index, I units:

$$I(p) = \sqrt{p}$$

(1) Sketch a graph of $p(t)$. Find out the upper limit on the population and estimate how long it will take to reach 90% of this limit.

(2) Sketch a graph of $I(p)$. What would be the pollution index when the population reaches its upper limit? What population would give a pollution index of 100? How long does it take for the pollution index to reach 100?

(3) What is the formula for the rate of population growth $p'(t)$ i.e. the number of new people per year? When is the rate of growth at its greatest?

(4) What is the formula for the rate at which the pollution index changes with the population, $I'(p)$? i.e. the number of units per person.

(5) What is the formula for the rate at which the pollution index grows with time, $I'(t)$ (i.e. the number of units per year). Show from your formulas that:

$$I'(t) = I'(p)\, p'(t)$$

This is the same as

$$\frac{dI}{dt} = \frac{dI}{dp}\frac{dp}{dt}$$

Population and Pollution

Help Sheet

(1) **PLOT** the graph of *p(t)* and use it to help you sketch the function. Note that the upper limit population is never reached. Use the algebraic formula to help you find the exact value of this upper limit. **SOLVE** an equation to help you find when the 90% level is reached but use figures from your graph to check.

(2) In answering the last question you will need to use the expressions for *I(p)* and *p(t)*.

(3) Use **DIFFERENTIATE** to help you find the rate of growth. If you **PLOT** a graph of the derived function *p′(t)* you can use it to help check your answer for the time when the rate is greatest.

(4) You might work this out without your computer algebra system!

(5) **SUBSTITUTE** to find *I* as a function of *t* and **DIFFERENTIATE** to find *I′(t)*. Then see if you get the same expression by multiplying together *I′(p)* and *p′(t)*.

Teaching notes

Background

The model used for population growth is the standard Verhulst-Pearl logistic curve applicable to population growth where there is a 'maximum' population which the environment can support. The full form of this model is $p(t) = \dfrac{NKe^{rt}}{K - N(1 - e^{rt})}$ where N is the initial population and K the 'maximum' population. It arises as the solution to the differential equation $\dfrac{dp}{dt} = r\left(1 - \dfrac{p}{K}\right)p$

The pollution index is a ficticious one – perhaps something this particular conservationist has devised for their own purposes!

Part of the aim of this activity is to give some experience of some aspects of modelling, it is also designed to reveal how the 'chain rule' for differentiation has applications and meanings beyond being a method for calculating derivatives.

Solutions

(1) The upper limit is 50,000 people. The exact time to 90% is 20 ln(3)/3 i.e. about 7.32 years. This can be found by solving the equation p(t) = 45000.

(2) The pollution index is √50000 i.e. about 224. A pollution index of 100 – a population of 10,000 – is reached after 10 ln(3/2)/3 years i.e. about 1.35 years. Population and pollution grow rapidly in the early phase as the growth is exponential.

(3) $\frac{30000e^{0.6t}}{e^{0.6t}+9} - \frac{30000e^{12t}}{(e^{0.6t}+9)^2}$. The greatest rate is approximately when $t = 3.6628$.

(4) $\frac{1}{2\sqrt{p}}$

(5) $\frac{30\sqrt{5e^{0.6t}}}{\sqrt{e^{0.6t}}+9} - \frac{30\sqrt{5e^{0.9t}}}{(e^{0.6t}+9)^{3/2}}$

Some algebraic re-arrangement and substitution is needed to show that the product of the answers to (3) and (4) gives the same expression as that in (5).

Max Cone

You wish to make a cone from a circular piece of paper. To do so you must cut out a wedge:

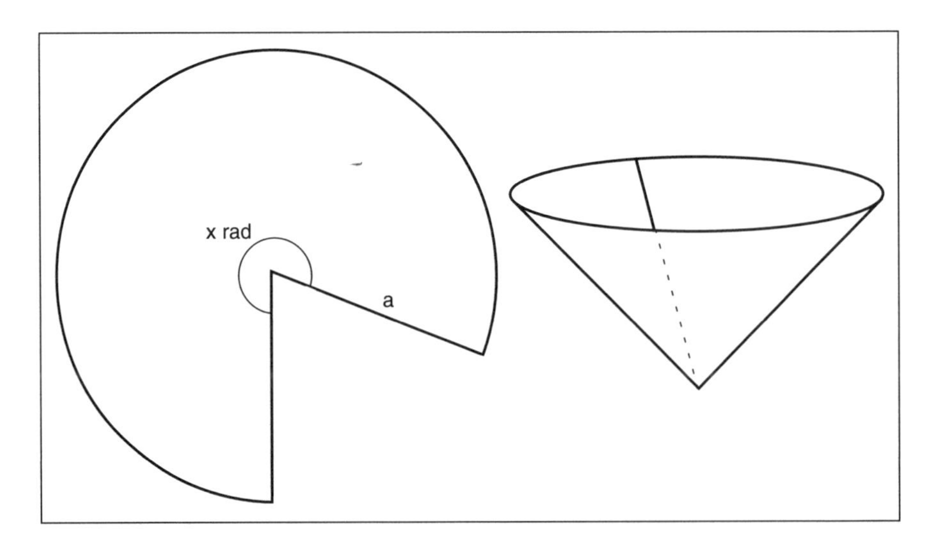

(1) What angle should you cut out so that the volume of the cone is a maximum?

Extension

(2) Now consider what would happen if you made a second cone from the remaining piece of paper you cut out. What angle should you cut out so that the *total* volume of the two cones is a maximum?

Max Cone

You will need to remember the formula for the volume of a cone of height, h, and radius, r is $V = \frac{1}{3}\pi r^2 h$

Also, remember that an angle of x radians in a circle of radius a makes an arc length of xa.

(1) There are two ways you can approach this problem depending on which variable you decide to use.

Method A:

Use the angle x radians as the independent variable. Then with the aid of your computer algebra system, work out expressions for:

- the circumference of the base of the cone
- the radius of the base of the cone
- the height of the cone
- the volume of the cone

Use **DIFFERENTIATE** to find the derivative and then **SOLVE** the equation $\frac{dv}{dx} = 0$ to find the maximum volume.

Method B:

Use the radius of the cone, r, as the independent variable. Find expressions for

- the height of the cone
- the volume of the cone

DIFFERENTIATE and **SOLVE** the equation $\frac{dv}{dr} = 0$ to find the maximum volume. Then **SUBSTITUTE** for r to obtain the final answer in terms of x, the angle.

Compare the two approaches above – try both ways.

(2) When tackling the two-cone problem remember that you must work with just one decision variable, such as x, and work out all other expressions in terms of it.

Teaching notes

Background

This is a straightforward optimisation problem though the expressions to be differentiated would be very lengthy without the assistance of computer algebra. Finding the initial expressions is moderately difficult though the diagram provided gives the key clues.

Solutions

It does not matter whether you use the small cut out wedge to make the cone or use the remaining bigger part of the circular piece of paper. The answer will be the same, though the algebra will be different. In the solution here, it is assumed that the larger part of the circular piece of paper is used to make the cone.

The first step is to link the variables. Taking a as the radius of the circular piece of paper and the angle cut out as x radians, the arc length of the wedge of paper is therefore xa. The circumference of the base of the cone, assuming you use the bigger part of the circle, is therefore $2a - xa$.

The radius of the cone, r, is therefore related to x by the formula $r = a - \frac{xa}{2\pi}$. The volume of the cone is therefore $V = \frac{1}{3}\pi r^2\sqrt{a^2 - r^2}$. The next step is to differentiate and solve the equation $V' = 0$. There are two approaches.

Either differentiate with respect to r to find the value of r which gives the maximum volume, and then find the corresponding value of x from the relationship above.

Or substitute the above expression for r into the formula for V and differentiate V with respect to x, thus finding the optimum value of x straight away.

The answer is that $z = 2\pi\left(1 - \sqrt{\frac{2}{3}}\right)$ radians, approximately 66°.

In the case of two cones, the second cone has a base with circumference xa and its radius and an expression for the total volume can be found in a similar manner to the above. The total volume is found by adding the two expressions, though care must be taken to keep just one variable in the entire expression, and not mix x's and r's.

The algebraic expressions are much more complicated than before, and it is much more difficult to obtain an answer. Depending on the way your particular computer algebra system works, you might find it easier to allow your CAS to work the answer in approximate numerical mode to find solutions. Or in emergency, use a graph to obtain an answer – but see the cautionary note below. There is a maximum at about 116.6 degrees and also at 243.4 degrees (each solution producing identical cones). There is a minimum at 180 degrees which might be mistaken for the solution if no-one checks carefully!

Discussion

The graph of the volume is useful to check answers.
The graph for the volume of two cones has an interesting flat feature which merits discussion. Depending on the scale chosen, the graph looks extremely flat and symmetrical and supports the notion that the maximum value occurs at $x = \pi$. This is misleading. A more careful look at the graph, with suitable use of the zoom facility reveals that in fact there are two symmetrical peaks with a slight minimum in between them.

Potential Student Difficulties

Students may encounter difficulty formulating the algebraic expressions initially. This process involves comparing variables on the diagram for the paper disk and the diagram for the cone. A fully labelled diagram is the only solution – so encourage students to take the trouble to produce one.

Some students may not notice that Pythagoras's theorem is needed to calculate the height of the cone – they may go off at a tangent and introduce trigonometry to find distances.

Students can become confused as to which variable to differentiate – is it x, r or even a? One confusion can be overcome by suggesting that the paper be of radius 10 cm.

When differentiating the expressions involving the variable x, some computer algebra systems will produce results involving modulus functions. This arises from the manipulation of the square roots. So the expressions may seem more complex than they really are. One way to reduce complexity is to substitute a numerical value for a. Expressions for the single cone problem are generally simpler using r as the decision variable.

Optimising Transport Costs

A transport company uses large lorries to transport goods by motorway. The manager wishes to know what would be the best average speed for drivers to maintain so as to keep costs down to a minimum. If a lorry goes too fast, then the petrol consumption will increase and thereby increase costs. On the other hand if a lorry goes too slow then the journey will take longer, and the costs will be increased as a result of paying extra wages.

Length of journey:	200 miles
Fuel consumption:	10 miles per gallon at 45 mph but reducing by 0.1 mpg for each extra mph over 45 mph.
Drivers wages:	£8 per hour
Fuel costs:	£2 per gallon

The cost of a journey is the fuel cost plus the wages cost

(1) What is the cost of a journey if the driver averages 50 mph?

(2) Work out a function which gives the cost of any journey in terms of the speed. Find out what speed gives the minimum cost, i.e. the optimum speed. Plot a graph of cost against speed showing the position of the minimum.

(3) Show whether the length of journey affects the optimum speed or the minimum cost.

(4) Show that the optimum speed depends on the **ratio** of wages to fuel costs.

(5) What ratio of wages to fuel costs gives an optimum speed of 50 mph?

(6) Investigate how the value of the fuel consumption affects the optimum speed. Write a summary of what you find in plain English for a manager to read.

Optimising Transport Costs

Help Sheet

(1) For a journey at 50 mph work out the cost yourself on your calculator.

(2) For a journey at *s* mph use your computer algebra system and **INPUT** expressions for

total wages cost (length of time for the journey multiplied by the hourly wages cost)
total fuel cost (number of gallons used multiplied by the fuel cost)
total cost (the sum of the other two)

SUBSTITUTE *s* with 50 to check that your expression is right.

Then find the optimum speed as follows:

DIFFERENTIATE the expression for the **total cost**.

Now **SOLVE** to find values for *s* which make the expression for the derivative equal to zero. Use **APPROXIMATE** to find the value of the optimum speed as a decimal.

Find the value of the minimum cost by **SUBSTITUTING** the *exact* value of the optimum speed into the original expression for total cost.

PLOT a graph of the original expression for **total cost**. Use the values you have just found to help you set the scale correctly.

(3) Go right back to the original expressions for **total wages cost** and **total fuel cost**. Make new expressions which have *d* instead of 200 for the distance travelled. Then go through exactly the same processes as before. Is *d* present in the final expression for optimum speed? Is it present in the final expression for minimum cost?

(4–6) For the other investigations, use variables for costs: *f* pounds per gallon of fuel, *w* pounds per hour wages. Then repeat the processes of finding the optimum speed. Similarly, the fuel consumption at 45 mph can be designated *c* mpg (rather than 10) and you can study the effect this has on the final results. Write a summary of what you find in plain English, for the manager to read.

Teaching notes

Background

This is a modelling activity where the modelling assumptions have been spelt out so that students can give attention to building an algebraic model and using calculus. However, the activity could easily lead to students finding out real data of their own based on particular types of lorry or a particular company.

To encourage students to interpret their mathematical results fully, you might suggest that at each stage they write a separate brief report to the manager of the transport company. This report should be in plain English, no mathematical formulas, but could give illustrative figures. The report should also assess the significance of the mathematics and include commonsense suggestions.

The model suggested for fuel consumption is a very simple one – a linear worsening of fuel consumption of 0.1 mpg for every extra mph over 45 mph. Calculations are only valid in the range 45 mph upwards and would not be reasonable at really high speeds. This is not as bad as it seems since real trucks will not be travelling at the extreme speeds! If you prefer you could replace the data on the worksheet by saying that fuel consumption is 10 mpg at 45 mph worsening to 8 mpg at 65 mph leaving the specific model to the student to find. Consumption should also worsen at low speeds, requiring a more complicated model.

Solutions

(1) For the 200 mile, four hour journey, the fuel consumption will be 9.5 mpg. This means that the **total wages cost** is £32, the **total fuel cost** is £42.11 giving a **total cost** of £74.11. The reason for asking this question is to enable students to become familiar with the method of calculation before going on to construct an algebraic version. This answer can then be used as a check on the expression developed on the computer algebra system.

(2) The total cost is $\frac{1600}{s} - \frac{4000}{s-145}$ and the value of s which corresponds to the minimum value is $\frac{145\sqrt{10}}{3} - \frac{290}{3}$, about 56 mph. The minimum cost is about £73. Note that the minimum cost at 56 mph is not very different from the cost of travelling at 50 mph. The **graph** indicates that a range of speeds would all be very close to the optimum value. This is a very important aspect of the interpretation of the mathematics. Though the optimum speed is 56 mph, you would not recommend this specific value to the manager of the transport company – a range of speeds will all give costs which are all very close to the minimum.

Ask students as a follow-up: what range of speeds will give a total cost below £75? Use the graph to give you an estimate. Use algebra to find the exact figures. The solution is in the approximate range 46 mph to 66 mph.

(3) Since the parameter d does not appear in the final expression for optimum speed, then you have proved that the distance does not affect the answer. The corresponding total cost will of course depend on distance, as longer journeys are bound to be more costly.

(4) Some computer algebra systems will not readily show the ratio of w/f and some experimentation will be needed to do this. It is probably simpler to carry out this stage of the algebra away from the computer:

$$s = \frac{145\sqrt{w}}{\sqrt{10}\sqrt{f} + \sqrt{w}} = \frac{145\dfrac{\sqrt{w}}{\sqrt{f}}}{\sqrt{10}\dfrac{\sqrt{f}}{\sqrt{f}} + \dfrac{\sqrt{w}}{\sqrt{f}}} = \frac{145\sqrt{\dfrac{w}{f}}}{\sqrt{10} + \sqrt{\dfrac{w}{f}}}$$

(5) The method for this is to solve the equation $\dfrac{145\sqrt{r}}{\sqrt{10} + \sqrt{r}} = 50$.

Students might solve it without using their CAS if they treat it as a linear equation for $\sqrt{r}$. Then they can use computer algebra to check they have the correct answer, rather than ask you, the teacher if they are correct. The ratio is 2.77.

(6) The expression for optimum speed replacing the figure of 10 mpg by c mpg is: $\dfrac{5\sqrt{w}\,(2c + 9)}{\sqrt{10}\sqrt{c} + \sqrt{w}}$. Students can use a range of numerical values or a graph to investigate the effect of changing values of c.

The last question is phrased as an investigation so that students can be encouraged to write a report to the manager of the transport company giving figures (but not algebraic formulas!) and explaining the effect of possible changes in consumption rate.

Students can also investigate the effect of changing the rate at which fuel consumption decreases with increased speed i.e. alter the value of 0.1. It is this particular model for how fuel consumption changes with speed which lies at the heart of the overall analysis.

This issue can be developed further into a discussion of the effect of lorry weight. Suppose you assume that the figure of 10 mpg corresponds to a weight of 30 tonnes. Fuel consumption might improve by 0.05 mpg for each tonne reduction in weight, and will worsen by the same amount for each extra tonne weight.

The Area Under a Curve

In this activity you will first approximate and then find an exact expression for the area under a function $y = f(x)$ between $x=a$ and $x=b$, $\int_a^b f(x)\,dx$.

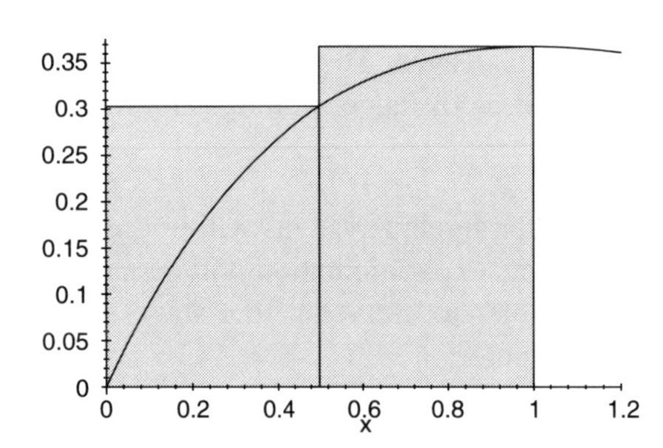

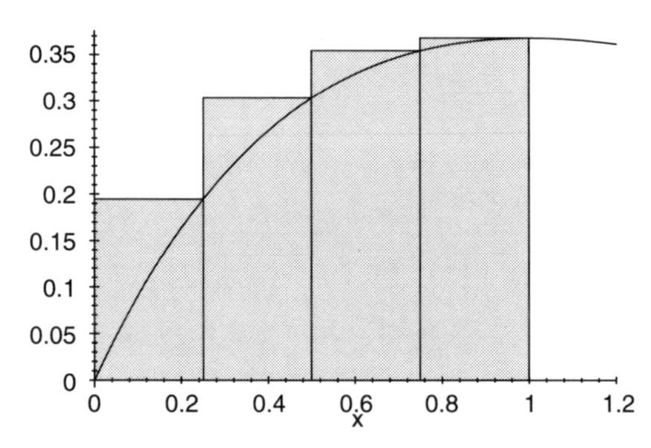

(1) Find an approximate value for $\int_0^1 xe^{-x}\,dx$ using two enclosing rectangles (this can be represented by the shaded area on the above). This calculation can be written as
$\frac{1}{2}\left[\left(\frac{1}{2}e^{-\frac{1}{2}}\right)+\left(\frac{1}{2}e^{-\frac{2}{2}}\right)\right]$ (this form is very useful for the generalisations to come).

(2) The area under the four rectangles gives a better approximation. Calculate this approximation and try to write it in this special form.

(3) Now try it for 10 rectangles and for 100 rectangles. This is too tedious to do by hand so get a computer algebra system to help you. The special form becomes cumbersome unless you write it in Σ notation.

(4) Now obtain an algebraic expression for the approximate area using n rectangles. Again, write it in this special form using Σ notation.

(5) Find the limit of this expression as n approaches infinity (∞).

(6) Now repeat steps (1) to (5) only this time for the area $\int_0^z xe^{-x}\,dx$.

(7) Now repeat steps (1) to (6) for functions other than $y = xe^{-x}$.

(8) Differentiate the functions you found in (6) and (7). What do you notice?

(2) The special form is

$$\frac{1}{4}\left[\left(\frac{1}{1}e^{-\frac{1}{4}}\right)+\left(\frac{2}{4}e^{-\frac{2}{4}}\right)+\left(\frac{3}{4}e^{-\frac{3}{4}}\right)+\left(\frac{4}{4}e^{-\frac{4}{4}}\right)\right].$$

(3) The special forms, in Σ notation, become $\frac{1}{10}\sum_{i=1}^{10}\frac{i}{10}e^{\frac{-i}{10}}$ and $\frac{1}{100}\sum_{i=1}^{100}\frac{i}{100}e^{\frac{-i}{100}}$. If you have problems putting these equations into the computer it may help to **INPUT** $\frac{i}{10}e^{\frac{-i}{10}}$ first, then ask your computer algebra system to **SUM** this (from i=1 to i=10) and then multiply it by $\frac{1}{10}$.

(5) You will need to use **LIMIT**. Once you have formed an expression, **SIMPLIFY** it. You may wish to rearrange this simplification, e.g. you may write $e^{-1}(e-2)$ as $\frac{e-2}{e}$.

(6) In the expressions above, the fractions $\frac{1}{2}, \frac{1}{4}, \frac{1}{10}, \frac{1}{100}$ and $\frac{1}{n}$ arose because we were dividing the x-axis, from $x = 0$ to $x = 1$, into 2, 4, 10, 100 and n equal parts. Here we need to divide the x-axis, from $x = 0$ to $x = z$, into 2, 4, 10, 100 and n equal parts.

(7) Choose your own functions like $f(x) = \sin(x)$ and $f(x) = x^3 - 2x^4$.

(8) Your functions are functions of z so you must differentiate with respect to z.

Teaching notes

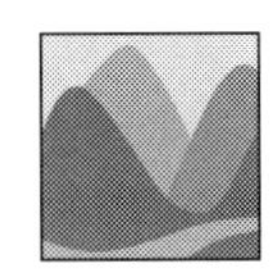

Background

This activity is designed to introduce the ideas of: i) integration as infinite summation, ii) the area function $F(z) = \int_{\alpha}^{z} f(x)\,dx$.

Computer algebra systems have been successfully used in introducing these concepts to students. There are indications too that this use aids students in seeing the use of integration in others areas such as volumes of revolution, centroids, lengths of arcs, etc.

The original version of this activity used the function $f(x) = x^2$ rather than $f(x) = xe^{-x}$. The latter was decided upon as it was not a function that could easily be done by hand. You may, however, want your students to begin with the easier function. In this case simply replace the 3 occurrences of xe^{-x} with x^2. The expression in (1) becomes $\frac{1}{2}\left[\left(\frac{1}{1}\right)^2 + \left(\frac{2}{2}\right)^2\right]$.

The difficulty of the concept, rather than the difficulty of using a computer algebra system, makes this an activity where more teacher supervision may be needed than for other activities in this book.

Enclosing rectangles (also called Riemann upper sums) have been used here. The activity can be amended so that lower rectangles are used instead.

It is recommended that students evaluate the approximations with two and four rectangles by hand. Hand calculations of approximations with more rectangles may be suggested if students do not see the pattern or to impress upon them the tedium of the process and thus the usefulness of the computer.

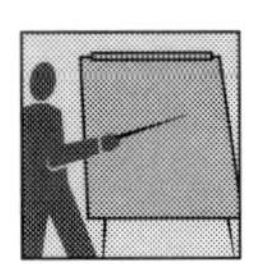

Solutions

Partial solutions to (1), (2), (3) and (5) are given in the HELP SHEET. Further to these:

(1) Area ≈ 0.3356. (2) Area ≈ 0.3050,

(3) Area using 10 and 100 rectangles ≈ 0.2818 and 0.2661, respectively;

(4) $\frac{1}{n}\sum_{i=1}^{n} \frac{i}{n} e^{\frac{-i}{n}}$

(5) $\lim_{n \to \infty} \frac{1}{n}\sum_{i=1}^{n} \frac{i}{n} e^{\frac{-i}{n}} = \frac{e-2}{e}$.

The solutions to the repeated steps in (6) are:

(1) $\frac{z}{2}\left[\left(\frac{z}{2}e^{-\frac{z}{2}}\right)+\left(\frac{2z}{2}e^{-\frac{2z}{2}}\right)\right]$

(2) $\frac{z}{4}\left[\left(\frac{z}{4}e^{-\frac{z}{4}}\right)+\left(\frac{2z}{4}e^{-\frac{2z}{4}}\right)+\left(\frac{3z}{4}e^{-\frac{3z}{4}}\right)+\left(\frac{4z}{4}e^{-\frac{4z}{4}}\right)\right]$

(3) $\frac{z}{10}\sum_{i=1}^{10}\frac{iz}{10}e^{\frac{-iz}{10}}$ and $\frac{z}{100}\sum_{i=1}^{100}\frac{iz}{100}e^{\frac{-iz}{100}}$

(4) $\frac{z}{n}\sum_{i=1}^{n}\frac{iz}{n}e^{\frac{-iz}{n}}$

(5) $\lim_{n\to\infty}\frac{z}{n}\sum_{i=1}^{n}\frac{iz}{n}e^{\frac{-iz}{n}}$.

Discussion

In (7) students are encouraged to repeat the activity for other functions. The wording was chosen so that students selected specific functions. Abstract functions, however, can be used in all systems up to the point where the limit is evaluated. This can save time (and reinforce theory) when steps (1) to (6) are repeated many times. In Derive, for example, you declare an 'empty function', *f*(*x*), by **INPUTTING** (use Author) *f*(*x*):= followed by 'enter' (↵). You can then generate $\lim_{n\to\infty}\frac{z}{n}\left(\sum_{i=1}^{n}f\left(\frac{iz}{n}\right)\right)$, specify *f*(*x*) and **SIMPLIFY** the expression.

Misconceptions

Conceptual difficulties not related to computer algebra surround this topic: limits and infinity take on a certain mysterious quality; infinite series are seen to "go on and on and never give an answer" regardless of whether they are convergent or divergent; students confuse sequences and series; focusing on the area under the curve as a number causes difficulties later when we wish to focus on the function generated by this process.

The form $\frac{1}{2}\left[\left(\frac{1}{2}e^{-\frac{1}{2}}\right)+\left(\frac{2}{2}e^{-\frac{2}{2}}\right)\right]$ is important for the later generalisation $\frac{1}{n}\sum_{i=1}^{n}\frac{i}{n}e^{\frac{-i}{n}}$. Some students will experience difficulty putting their calculation in this form and (correctly) write something like $\frac{1}{2}e^{-1}+\frac{1}{2}\left(\frac{1}{2}e^{-\frac{1}{2}}\right)$.

In theory computer algebra systems should not make a distinction between **SIMPLIFYING** $\lim_{n\to\infty}\frac{1}{n}\sum_{i=1}^{n}\frac{i}{n}e^{\frac{-i}{n}}$ and **SIMPLIFYING** $\lim_{n\to\infty}\frac{n-e^{\frac{1}{n}}(n-e+1)}{en^2\left(e^{\frac{1}{n}}-1\right)^2}$. The authors have found, however, bugs in versions of some computer algebra systems so that they do the latter but not the former. It is a good idea for you to try this before your students do it.

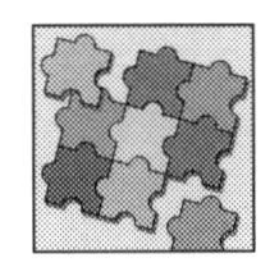

Extensions

- An additional and related task is to plot both functions and to compare them in much the same manner as is encouraged in the Activity Worksheet *Function and derivative: visualisation.*

- Instead of always taking integrals starting with 0 start with another number (or a parameter, a).

- Instead of taking $z > a$ in $\int_a^z f(x)\,dx$, take $z < a$.

- Plot the functions $\int_a^z f(x)\,dx$ for different values of a (you will get a family of curves $F(z) + c$).

Find the exact area of the shaded regions below.

(1) $y = x^3 - 4x$
$y = 0$

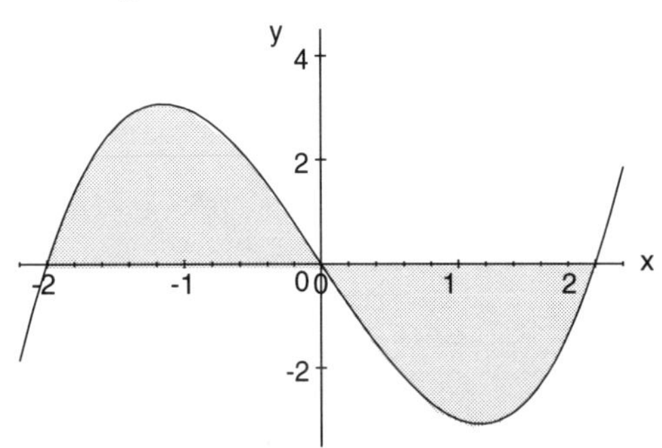

(2) $y = 4 - x^2$
$y = 2$

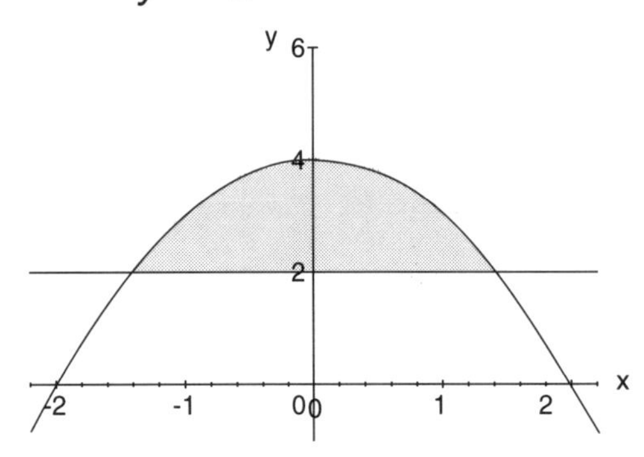

(3) $y = x^2 - x - 2$
$y = x^3 - 2x^2 - x + 2$

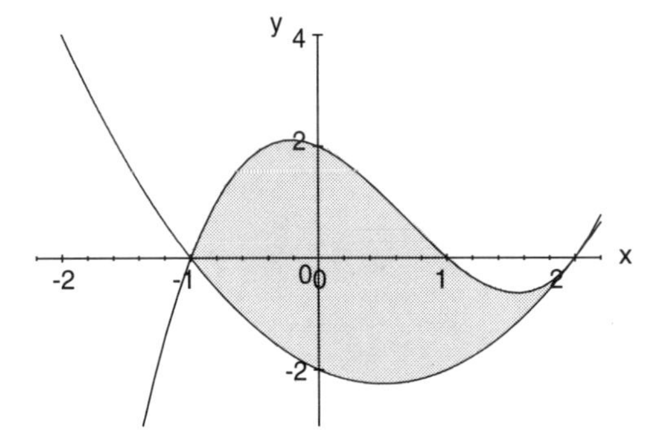

(4) $x^2 + y^2 = 1$

$x^2 + \frac{y^2}{4} = 1$

$\frac{x^2}{4} + y^2 = 1$

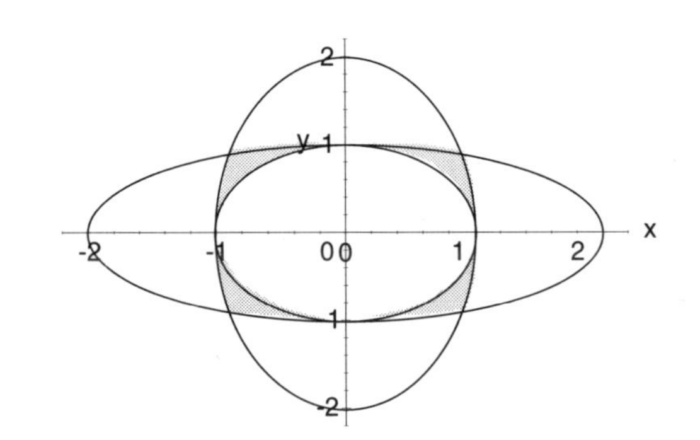

Enclosed Areas

Help Sheet

Before starting each problem roughly estimate the shaded areas using rectangles or triangles.

You can find the area under a curve between two points by integrating the equation of the curve between those two points. The integral gives the area between a curve and the x axis between the two x values chosen as limits of integration. If the area is above the x axis it is positive and if it is beneath the x axis it is negative. When you have two curves, the area between them can be found by subtracting the two integrals. However, this only works if the areas are both positive or both negative.

(1) Treat the areas above and below the x axis separately.

(2) **SOLVE** to find the points of intersection of the straight line $y = 2$ and the curve $y = 4 - x^2$. Decide which areas you need to calculate, in order to find the shaded area, before you continue.

(3) Again, split the shaded area into sections which you can calculate. Remember, if the area is below the x axis it will be negative.

(4) This is not easy! Use the symmetry of the area to minimise calculation. You will need to write the equations in the form $y = f(x)$, before you **INTEGRATE**, you will also need to **SOLVE** to find the points of intersection of the two ellipses.

Teaching notes

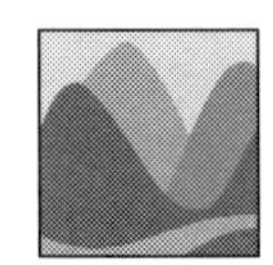

Background

This activity requires the knowledge that the area under a curve can be found from definite integration. Question (4) is a much more demanding activity that is designed to stretch the more able, you may wish to blank this out. Also, blanking out the equations of the curves will add another level of difficulty for brighter students.

Solutions

(1)

$$\int_{-2}^{0} (x^3 - 4)\, x\, dx + \left| \int_{0}^{2} x^3 - 4x\, dx \right| = \int_{-2}^{2} \left| x^3 - 4x \right| dx = 8$$

To avoid using the modulus function on the CAS, students can also integrate form 0 to 2, giving the answer of –4, and take the negative value 4. Owing to the symmetry of the function the answer is double this figure, i.e. 8.

(2) Equate and solve the equations $y = 2$ & $y = 4 - x^2$ and we get $x = -\sqrt{2}, x = \sqrt{2}$. The area under the line is $4\sqrt{2}$. The shaded area can be found by subtracting the area under the line from that under the curve.

Hence the shaded area is

$$\int_{-\sqrt{2}}^{\sqrt{2}} (4 - x^2)\, dx - 4\sqrt{2} = \frac{8\sqrt{2}}{3}$$

An alternative approach is to use the principle that the area between the two functions $f(x)$ and $g(x)$ may be found by integrating the difference $f(x)$-$g(x)$. In this case

$$\int_{-\sqrt{2}}^{\sqrt{2}} (4 - x^2 - 2)\, dx = \frac{8\sqrt{2}}{3}$$

(3) By considering the different sections the area is

$$\int_{-1}^{2} (x^3 - 2x^2 - x + 2)dx + \left| \int_{-1}^{2} x^2 - x - 2\, dx \right| = \frac{27}{4}$$

Again, alternative approaches that avoid the use of modulus signs can be used.

(4) Due to symmetry only one shaded section needs calculating. Firstly find the x co-ordinates of the point of intersection of the two ellipses in the first quadrant. The co-ordinates are $\left(\frac{2\sqrt{5}}{5}, \frac{2\sqrt{5}}{5} \right)$.

Rearrange the equations of the ellipses into the form $y=f(x)$, to give

$$y = \sqrt{1-x^2}$$

$$y = 2\sqrt{1-x^2} \qquad (y_1)$$

$$y = \sqrt{1-\frac{x^2}{4}} \quad \text{or} \quad y = \frac{\sqrt{4-x^2}}{2} \qquad (y_2)$$

Taking the positive square root indicates that we are working in the region $y > 0$.

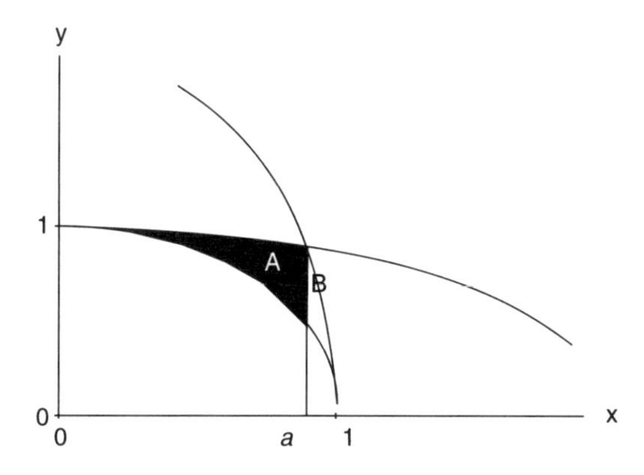

By consideration of the above diagram the area of **A** is

$$\int_0^a y_2 dx - \int_0^a \sqrt{1-x^2}dx = \frac{3\tan^{-1}\left(\frac{1}{2}\right)}{2} - \frac{5\pi-4}{20}$$

and the area of **B** is

$$\int_a^1 y_1 dx - \int_a^1 \sqrt{1-x^2}dx = \frac{\tan^{-1}\left(\frac{1}{2}\right)}{2} - \frac{1}{5}$$

where $a = \frac{2\sqrt{5}}{5}$. This gives the area of **A** + **B** as

$$2\tan^{-1}\left(\frac{1}{2}\right) - \frac{\pi}{4}$$

so the total shaded area is $4\left(2\tan^{-1}\left(\frac{1}{2}\right)-\frac{\pi}{4}\right) = 8\tan^{-1}\left(\frac{1}{2}\right) - \pi$

which is approximately 0.567587.

A Function whose Derivative is Itself

The functions x^3 and sinx differentiate to $3x^2$ and cos$x$, respectively. These functions and their derivatives look quite different and produce different graphs. Is it possible to find a function whose derivative looks and graphs exactly the same?

(1) Graph the curve $y = 2^x$ and use your computer algebra system to differentiate 2^x and plot the answer. Write down how the graph of this derivative compares with that of the original function $y = 2^x$.

(2) Repeat (1) with $y = 3^x$. Is this any better?

(3) Find the number a which will make the function $y = a^x$ differentiate to itself! The symbol for this special value is e.

(4) Using *your* value for e, investigate what happens to the derivatives of $y = 2^x$, $y = 3e^x$, $y = 4e^x$, etc. Hence, write down the derivative of $y = Ae^x$, where A is any number.

(5) Find a function whose derivative is exactly twice the original function.

(6) Now find functions whose derivatives are exactly three, four, five etc. times the original function.

(3) The graph of the derivative of 2^x is below the graph of $y = 2^x$, the graph of the derivative of 3^x is above the graph of $y = 3^x$. Choose a number between 2 and 3, say 2.5, and determine whether the graph of the derivative of 2.5^x is above or below the graph of $y = 2.5^x$. Continue like this by trial and improvement until you get an accurate match. Use **ZOOM** to achieve higher accuracy.

(5) Find the new value for a for which the derivative of a^x is twice the original function.The following trick may help you with this problem:

To see if a function is twice the original, first use your CAS to calculate the ratio of the derivative of the function divided by the original function. Then **PLOT** the graph of this ratio. If the derivative is really twice the original, then you would see a straight line y=2. In general, if the derivative is not a multiple of the original function then the graph of their ratio will not be straight and horizontal.

(6) Find the values of a which make the derivative of a^x three times, four times, etc. the original function. When you have finished several of these see if you can spot a pattern in the values of a – how are these related to e?

Teaching notes

Background

This is an introduction to the function e^x and the solution of differential equations of the type $\frac{dy}{dx} = Ay$. Any CAS will have a special symbol for e, you may wish to mention this after students have found their approximation to it. For the further exercises students may wish to use this instead of their approximation. It may not be necessary to simplify any of the derivatives as most CASs will plot expressions such as $\frac{d}{dx}2^x$, its simplification to $\ln(2)2^x$ would be unhelpful at this stage. It would be prudent to clear the graphs each time you start a new question, in an investigation like this the plot area can become cluttered and individual plots difficult to see.

Solutions

(1) It has a similar shape, but it is below the original curve.

(2) Again a similar shape, but this time the graph of the derivative is above and much closer to the original curve.

(3) The number we are looking for is between 2 and 3, so try 2.5 the mid point of 2 and 3, if the graph of the derivative is below the original function then the number must be in the interval [2.5,3]. Repeating this algorithm will eventually lead to the derivative plotting on top of the original curve. At a scale of x=1, y=1 it will appear to plot over the top at a=2.75. Below is a table of values of a that the derivative appears to plot over the original function at different scales.

a	*Scale x*	*Scale y*
2.75	1	1
2.718	0.2	0.2
2.7183	0.0001	0.001

(4) In each case the graph of the derivative plots over the original function, therefore the derivative of $y = Ae^x$ is Ae^x.

(5) The students may well have difficulty in getting started here, so a little nudging in the right direction may be necessary. Also, graphically it may not be obvious that one function is always, say, twice another function. So plotting the quotient of the derivative and the original function will give a horizontal straight line y=2, if the derivative is twice the original function.

Answer $y = 7.389^x$, which is $y = (e^2)^x$

(6) The values of a which makes the derivative of a^x equal 3, 4 & 5 times a^x are approximately 20, 54.5 & 148.5 respectively. Which are e^3, e^4, & e^5 respectively. Some students may need gentle pushes in the right direction.

We lead to the general result that, if

$$y = e^{ax} \text{ then } \frac{dy}{dx} = ae^{ax} = ay.$$

Extensions

In activity 1 we found a function that differentiates to itself, if this function is y then this relationship can be written as $\frac{dy}{dx} = y$, this is called a differential equation. Find a possible solution to the differential equation $\frac{dy}{dx} = y$.

Find an expression for all possible solutions to the differential equation $\frac{dy}{dx} = 3y$.

Answer $y = Ae^{3x}$, where A is any real number.

Examine, or at least imagine, a wine glass. The diagram shows three common designs. Now focus on the bowl (the part that holds the wine). Imagine it is turned 90°. Then we could generate the volume by revolving a function $y = f(x)$ about the x-axis clockwise.

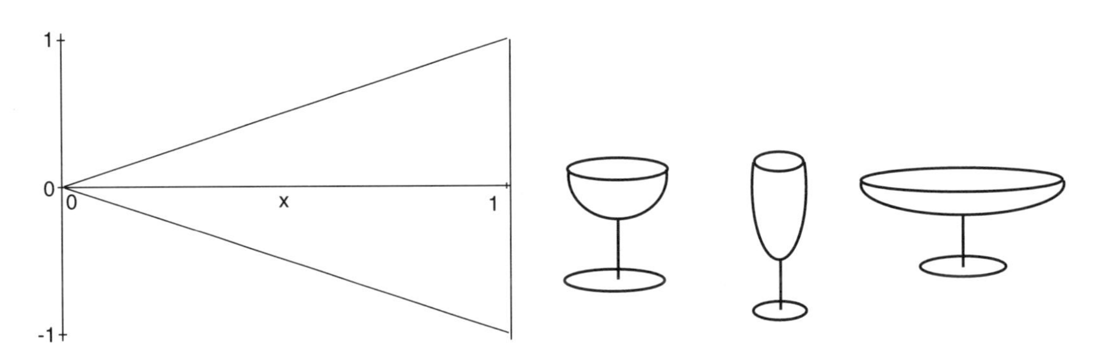

In this activity you are asked to design a wine glass bowl to hold about 150 ml[1] of liquid (anything from 130 to 170 ml will do). The height of the bowl should be between 50 and 100 mm. You should try out different functions and come up with various shapes that fit the specifications and look attractive.

(1) Design a conical wine glass bowl of height 100 mm which will hold 150ml of liquid. Do this imagining the glass is turned 90° clockwise, as in the figure above, and representing the equation of the side by $f(x) = mx$, i.e. $y = mx$. Rotate this equation about the x-axis (between $x = 0$ and $x = 100$) to obtain a cone of volume 150ml. Solve the resulting equation to obtain m.

(2) **Other shapes to investigate.** The following functions are possible.

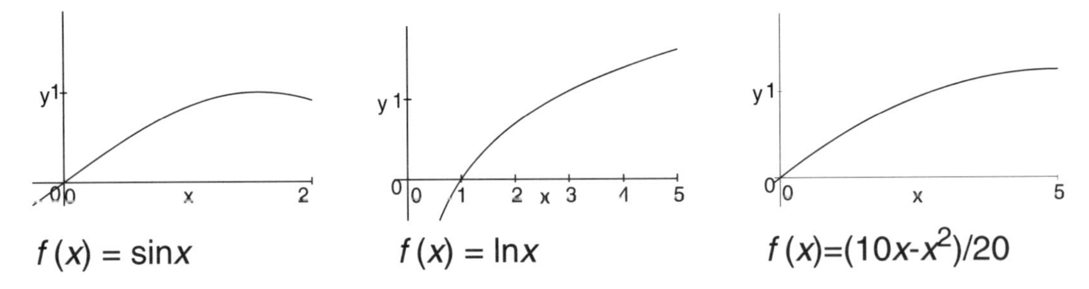

$f(x) = \sin x$ $f(x) = \ln x$ $f(x) = (10x - x^2)/20$

You must, of course, transform these functions to meet the wine glass specifications. You could transform, say, $f(x) = \sin x$, using parameters k and l, $f(x) = k \sin(lx)$ (k and l affect the y and x-axes respectively –experiment with different values of k and l and see for yourself).

For each function (and others of your own choosing) revolve it about the x-axis and find the volume using the formula $\pi\int_a^b (f(x))^2 dx$. Make adjustments to k, l and $f(x)$ so that the volume and height are within the specified limits.

1. Recall that 1 ml is 1 cubic centimetre, i.e. 10^3 mm^3 =1000 mm^3.

Recall that if a function, $f(x)$, is rotated (revolved) about the x-axis between $x=a$ and $x=b$. The volume of the resulting solid is $\pi\int_a^b (f(x))^2 dx$

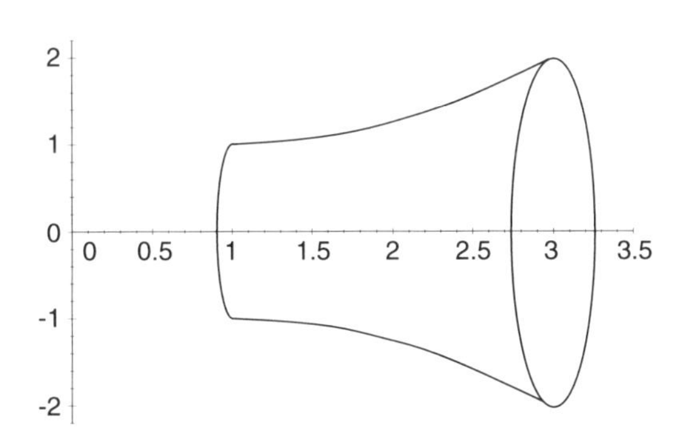

(1) The equation of a straight line through the origin is $y = mx$.

So, $V = \pi\int_0^{100} (mx)^2 dx$. If we evaluate this integral we obtain $\frac{\pi m^2 10^6}{3}$ mm^3. But V = 150 ml = 1.5×10^5 mm^3. Thus, $\frac{\pi m^2 10^6}{3} = 1.5 \times 10^5$. **SOLVING** for m, and ignoring negative values, we get $m = \frac{3\sqrt{5}}{10\sqrt{\pi}} \approx 0.3785$.

(2) We will try one possible design with the first function. Assume the x-axis represents mm. The function as it stands gives a bowl about 2 mm high. We need this to be 50 to 100 mm. $f(x) = \sin\left(\frac{x}{50}\right)$ will give us a bowl 100 mm high. The volume, however, is

$\pi\int_0^{100}\left(\sin\left(\frac{x}{50}\right)\right)^2 dx = 50\pi - \frac{25\pi\sin 4}{2} \approx 187$. Now we must be very careful about units here. 187 represents 187 mm^3 = 0.187 ml. Clearly this is too small!! A possible strategy is to work with $f(x) = k\sin\left(\frac{x}{50}\right)$ and equate this to, say, 150 ml (=150 000 mm^3) to find k.

$$\pi\int_0^{100}\left(k\sin\left(\frac{x}{50}\right)\right)^2 dx = k^2\pi\left(50 - \frac{25\sin 4}{2}\right) = 150\ 000 \Rightarrow k \approx 28.3$$

Note that k here is a stretch factor along the y-axis. The width of this wine glass at its widest point is thus $2k$ mm.

The calculations in (2) represent only one of many possible designs. You may, as a first step, like to repeat the workings with a bowl height of 80 mm (you should find that $k \approx 31.7$) before you go on to more creative designs of your own.

Teaching notes

Background

This activity is conceived of as a short project for students after they have covered *Volumes of Revolution.* It allows for some creativity and plenty of reinforcement of the basic ideas.

Solutions

As the specifications are within limits students are likely to come up with many different designs. This could provide a time problem for the teacher who is checking their work. Let the computer algebra system ease your marking load here. Below are the commands we used to obtain $k \approx 28.3$ in part (2) of the Help Sheet. This routine, with different functions, limits of integration, etc. can be used repeatedly.

- **INPUT** the equation of the function
- **INTEGRATE** and specify the limits of integration
- Set this integral equal to 150 000 (or whatever)
- **SOLVE** (for k in this instance)
- **APPROXIMATE**

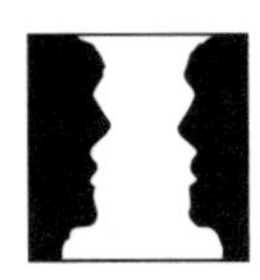

Discussion

Some teachers may prefer to rotate the curve about the y-axis. This avoids rotating the bowl 90° and gets away from always doing volumes of revolution about the x-axis.

Computer algebra systems will be very useful in performing the integrations and for plotting the functions. However, it is extremely difficult to program them to display the resulting solid.

We have worked with ml here (which is consistent with the SI system). Some wine glasses specify ml, others specify cl. You may wish to use cl. The specifications in the worksheet can be easily changed.

Misconceptions

The main problem is likely to be with units. The author of this worksheet actually made a number of 'silly' mistakes himself when he first tried this. It is recommended that you try an example in order to anticipate potential student errors.

It is easy to incorrectly write, say, $\pi k \int_0^{100} \left(\sin\left(\frac{x}{50} \right) \right)^2 dx$ instead of

$\pi \int_0^{100} \left(k \sin\left(\frac{x}{50} \right) \right)^2 dx.$

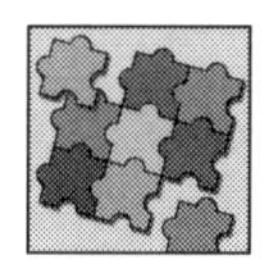

Extensions

In (2) in the Help Sheet the width at the widest point is $2k$. What is the width at the rim?

If you have covered the topic 'centre of mass', then find the centre of mass of the bowls you have designed in this activity. In terms of stability, is a high or low centre of mass better?

Activity Worksheet

The Limit of a Sequence

This activity is based around different sequences of numbers s_n where n is a natural number (1, 2, 3, 4, ...). For example the sequence 2, 4, 6, 8, ... is described by the formula $s_n = 2n$ so that $s_1 = 2$, $s_2 = 4$, etc.

8 sequences are given below. This activity will ask you to calculate some terms, explore their graphs, find out whether or not they have limits and thus classify their behaviour.

(1) For each of the sequences below evaluate the 1st, 2nd, 10th, 20th, 100th and 1000th terms.

$a_n = 3 + \frac{1}{10^n}$	$b_n = \frac{\ln(n)}{n}$	$c_n = \frac{n^2 - n + 1}{n^2 + n + 1}$
$d_n = \frac{n}{\ln(n)}$	$e_n = \cos(\pi n)$	$f_n = n \sin(n)$
$g_n = \frac{\sin(n)}{n}$	$h_n = \left[\frac{20}{n}\right]$ (this means the *integer* value of $\frac{20}{n}$)	

(2) Can you classify their behaviour? What will the 1 000 000th term be?

(3) Now look at the graph of each one. Move around and **ZOOM** in and out.

(4) Now explore the behaviour of each sequence as n approaches infinity (∞) and explore whether or not there is a limit.

(5) Have you any changes to make in your classification of their behaviour?

The Limit of a Sequence

Help Sheet

In each of these sequences n is a natural number e.g. 1, 2, 3, 4, ...

(1) To find the 1st term of $a_n = 3 + \frac{1}{10^n}$ simply put 1 in, e.g. $3 + \frac{1}{10^1} = 3.1$. You can do this by **SUBSTITUTION** on your computer algebra system but it might be faster doing it on your calculator.

(2) Ideally the classification will be your own and you can compare your classification with others in the class. However, if you not sure what is meant, then group together those sequences that, as n gets larger and larger: *settle down, oscillate, get larger without bound, etc.*

(3) The graphs will probably appear as curves whereas you really want the graph of points at each whole number value of n. Working with the curves is fine as long as you remember that you are really only interested in the points up from whole numbers.

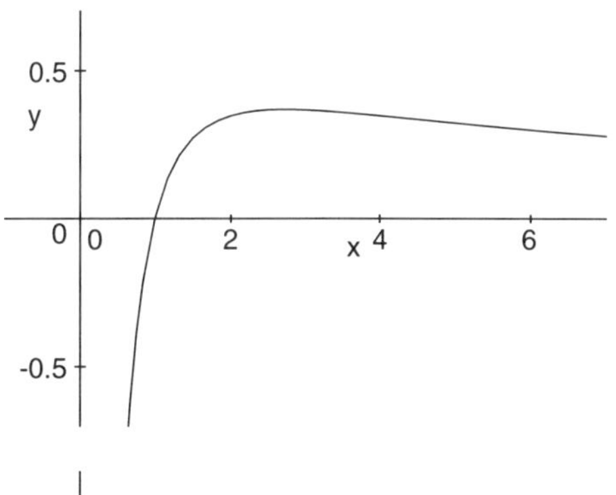

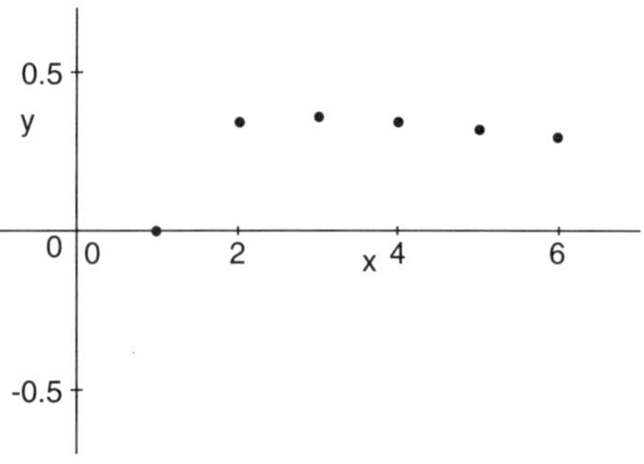

> $[x]$ is the greatest integer less than or equal to x, e.g. $[3.7] = 3$, $[-3.7] = -4$.
>
> Most computer algebra systems use the term Floor(x).

(4) When you get the expression, say, $\lim_{n \to \infty}\left(\frac{\ln(n)}{n}\right)$ remember to **SIMPLIFY** it. You will get some symbols such as ∞ or ? or even a comment like "infinity" or "undefined". Look at carefully at these sequences. Their behaviour as n gets very large should explain what is happening.

Teaching notes

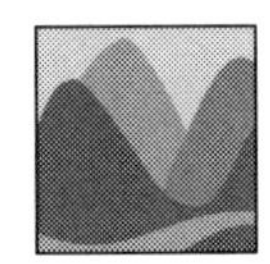

Background

This activity is based on an idea from the ATM Activity Book "These have worked for us at A-level".

Solutions

The sequences illustrate many possible behaviours of sequences.

$\{a_n\}$	approaches a limit from above	the limit is 3.
$\{b_n\}$	approaches a limit from above	the limit is 0.
$\{c_n\}$	approaches a limit from below	the limit is 1.
$\{d_n\}$	is an increasing sequence	it has no limit.
$\{e_n\}$	is a finitely oscillating sequence	it has no limit.
$\{f_n\}$	is an infinitely oscillating sequence	it has no limit.
$\{g_n\}$	is an oscillating sequence which tends to a limit	the limit is 0.
$\{h_n\}$	is a series that *reaches* its limit	the limit is 0.

Misconceptions

It is important that all the sequences are examined closely. Students tend to think that:

- Oscillations prevent a sequence approaching a limit (because it moves away again!).
- A sequence that attains or passes through a limit does not approach it.

Try to use all of the phrases *approaches, tends to, converges* and *has a limit* in your discussion. Often the everyday meaning of these words obscures the fact that they are equivalent in mathematics. Sometimes the phrases mean different things in numeric and graphic contexts e.g. "How can numbers converge? Only lines (or curves) can converge."

The hand calculations, graphs and computer algebra components are all seen as important.

When students do hand calculations they tend to focus on the *process* of getting near a limit when finding specific terms and fail to see that there is a limit *object* at the end. Using the computer algebra system should change their focus from the process to the object.

The graphs should help your students obtain an overall picture of what is happening "in the limit".

It may be an advantage to get the students to sketch the graphs on paper. It is common to find students losing track of their graph when they **ZOOM** in or out. Hand calculations can help them anticipate where the graph should be.

When using the computer algebra system it will probably be useful to first INPUT the sequence, e.g. $\frac{\ln(n)}{n}$, and then manipulate it, e.g. tell the computer algebra system to find the limit. These systems need to be told what the limit variable is, e.g. n, and what value this is approaching, e.g. ∞.

Some computer algebra systems cannot handle limits algebraically. At the time of writing *Theorist* can only estimate the limit of a function at a point graphically or numerically.

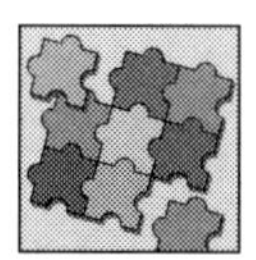

Extensions

- Generate other sequences with similar behaviours to $\{a_n\}$ ... $\{h_n\}$.
- Examine the behaviour of the *series* of positive terms of $\{a_n\}$ ... $\{h_n\}$.

Activity Worksheet

Visualising Taylor Approximations

Recall that a *polynomial* is a function ($f(x) = ...$) that can be written in terms of positive powers of x, e.g. $f(x) = x^3 - 3x^2 + 5$. Many functions, like $f(x) = 1/x$ or $f(x) = \sin x$, are not polynomials but can be *approximated* by polynomials, e.g. $\sin x \approx x - \frac{x^3}{6} + \frac{x^5}{120}$ when x is near to 0. This is called the Taylor[1] approximation of degree 5 and centre $x = 0$ (or order 5 about the point $x = 0$). The purpose of this activity is for you to get a feel for the relationship between the graphs of a non-polynomial function and its polynomial approximation.

(1) Take a non-polynomial function (in this example we'll use $f(x) = \frac{1}{1+x^2}$). Obtain the Taylor approximation of degree 3 about the centre $x = 0$.

(2) Sketch the function and the Taylor approximation on the interval (–2,2). Between which x values do the two functions appear to be the same? Make a note of these.

(3) Now **ZOOM** into the region where the curves appear to be the same. Do they still look the same? Describe differences and similarities between these two views.

(4) Now 'play around' by going back to the original function and change the degree or the centre or both. Sketch the new approximation. **ZOOM** in if you want to. Keep doing this until you can clearly see what visual effect the order and the centre has on the approximating polynomial.

(5) Now choose another non-polynomial function and repeat steps (1) to (4). Some functions are suggested below but it would be a good idea to make up some of your own. Remember that the degree should be a positive whole number. Sometimes a centre you pick may not be suitable (looking at your function should tell you why).

Some other functions to try:

$$f(x) = \frac{x}{1+x^2}, \quad f(x) = \cos x, \quad f(x) = e^x,$$

$$f(x) = \arctan x, \quad f(x) = \ln(1 + x)$$

1. After the British mathematician Brook Taylor (1685–1731).

Visualising Taylor Approximations

Help Sheet

Much depends on the order of the approximation and the centre of the approximation. The following examples illustrate these two factors.

$\sin x \approx x$ (degree 1, centre 0) $\quad \sin x \approx x - \frac{x^3}{3!} + \frac{x^5}{5!}$ (degree 5, centre 0)

$\sin x \approx \frac{1}{384}(16x^4 - 32\pi x^3 + 24x^2(\pi^2 - 8) + 8\pi x(24 - \pi^2) +$

$+\pi^4 - 48\pi^2 + 384)$ (degree 5, centre $\pi/2$)

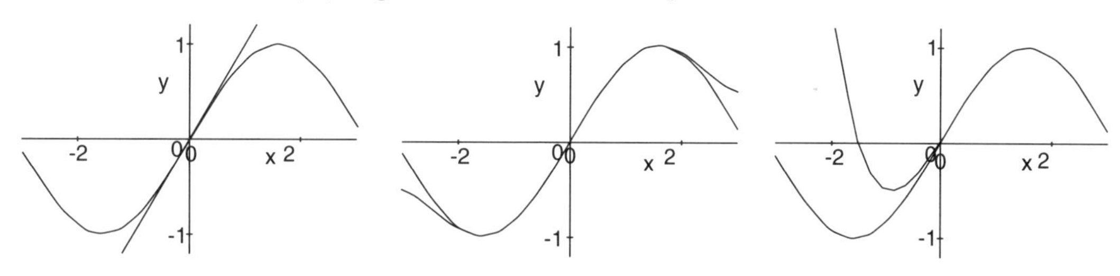

A number of things may appear to go wrong. Here are some of them:

- Your graphs may not appear. The most likely problem is that they are outside of your viewing area. Try ZOOMING out along both the *x* and *y*-axes.
- You appear to get only one graph. In this case the most likely explanation is that your Taylor approximation is so good that it has been plotted on top of the original function. You should be able to see the two graphs if you **ZOOM** out along the *x*-axis.
- Your Taylor approximation of, say, sin*x*, is not shown as a polynomial, e.g. you may get TAYLOR(sin*x*,*x*,0,5). In this case tell the computer algebra system to **SIMPLIFY** the expression.
- Your Taylor approximation doesn't appear in the form you expected. In this case the computer algebra system has probably simply given you an algebraic rearrangement of the function, e.g.

  $\frac{x^4}{4!} - \frac{x^2}{2!} + 1$ instead of $1 - \frac{x^2}{2!} + \frac{x^4}{4!}$.

Teaching notes

Background

This activity may come before or after the formal derivation of Taylor approximations. Although there is a strong case that the formal derivation should come first so that students know what is happening, the activity can be used to give students a visual overview that helps them to understand the formal derivation.

Solutions

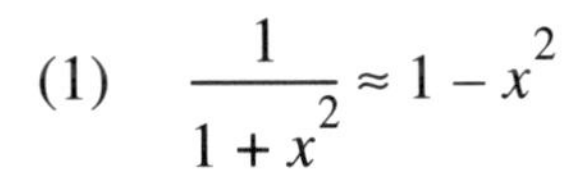

(1) $\dfrac{1}{1+x^2} \approx 1 - x^2$

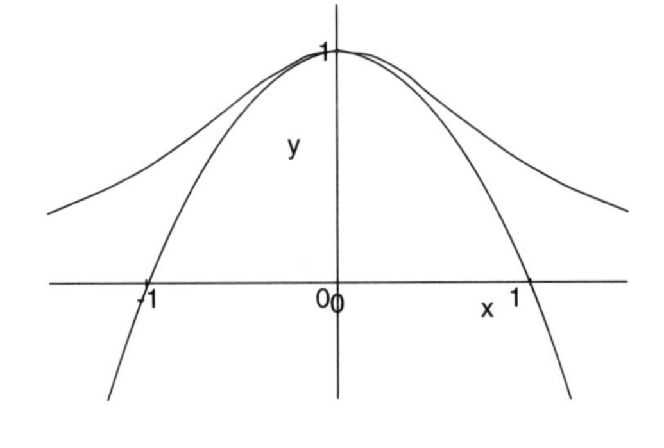

(2) Solutions will clearly vary according to the degree and centre of the approximation.

Algebraic solutions (degree 5, centre 0) of the functions at the bottom of the Activity sheet are:

$$\frac{x}{1+x^2} \approx x - x^3 + x^5 \qquad \cos x \approx 1 - \frac{x^2}{2!} + \frac{x^4}{4!} \qquad e^x \approx 1 + x + \frac{x^2}{2!} + \frac{x^3}{3!} + \frac{x^4}{4!} + \frac{x^5}{5!}$$

$$\arctan x \approx - \frac{x^3}{3} + \frac{x^5}{5},\ |x| \le 1 \qquad \ln(1+x) \approx x - \frac{x^2}{2} + \frac{x^3}{3} + \frac{x^4}{4} + \frac{x^5}{5},\ -1 < x \le 1$$

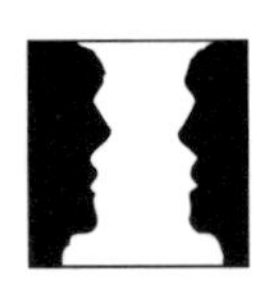

Discussion

An interesting discussion point for the classroom is that the computer algebra system usually uses Taylor approximations when it evaluates non-polynomial functions. This can cause some amusement in that we are actually approximating approximations!

Misconceptions

Steps (3) and (4) can present technical problems for students who are not very familiar with the computer algebra system they are using.

Two common things that 'go wrong' which are not included on the HELP SHEET (because of the complications involved in explaining them briefly) are:

- The function is not defined at the chosen centre, e.g. $\ln x$ at $x=0$ and $\dfrac{1}{1+x}$ at $x=-1$.

- Attempting to approximate outside of the interval of convergence, e.g. $\arctan x$ and $\ln(1+x)$.

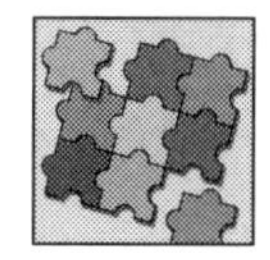

Extensions

- Explore the interval of convergence.

- Plot the graphs of $f(x) = \sin x$ and $g(x) = x(\cos x)^{\frac{1}{3}}$. Note that they agree rather well on the interval (–1, 1) and very poorly outside this interval. Use your computer algebra system to generate the 11th degree Taylor polynomial for these functions about $x = 0$. What do you notice about these polynomials? Does this explain why they seem to agree on (–1, 1) and disagree elsewhere?[1]

1. The example comes courtesy of Carl Leinbach (1991), *Calculus Laboratories Using DERIVE*, Wadsworth Pub. Co., Belmont, Ca. Carl, in turn, attributes this to Jerry Uhl of the University of Illinois.

Activity Worksheet

Visualising Matrix Transformations

A matrix such as $\begin{pmatrix} 5 & -3 \\ 1 & 1 \end{pmatrix}$ represents a transformation that can act on vectors. For example, the vector $\begin{pmatrix} 3 \\ 1 \end{pmatrix}$ is transformed to $\begin{pmatrix} 5 & -3 \\ 1 & 1 \end{pmatrix}\begin{pmatrix} 3 \\ 3 \end{pmatrix} = \begin{pmatrix} 12 \\ 4 \end{pmatrix}$. You will notice that the direction of this vector is unchanged, even though its length has been multiplied by 4. So $\begin{pmatrix} 3 \\ 1 \end{pmatrix}$ is called an *eigenvector* and the multiple 4 is called an *eigenvalue* for the matrix $\begin{pmatrix} 5 & -3 \\ 1 & 1 \end{pmatrix}$.

The vectors $\begin{pmatrix} 3 \\ 1 \end{pmatrix}$ and $\begin{pmatrix} 12 \\ 4 \end{pmatrix}$ can be represented graphically, as can the line through both endpoints (in this case $y = \frac{x}{3}$) .

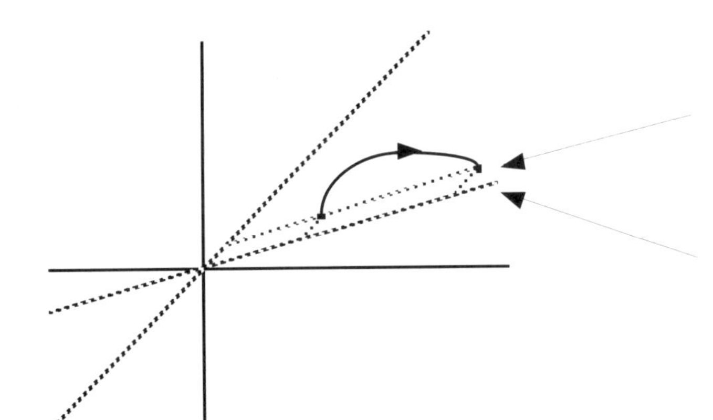

(1) Find another eigenvalue and eigenvector for this matrix, possibly by using trial and improvement. Illustrate these vectors as above.

(2) Explore other vectors. How do their transformations relate to the two eigenvectors found above?

(3) Can you find a way to determine the eigenvalues and eigenvectors more quickly by using an equation that can then be solved?

(4) Use the methods above to explore these or other matrices. What do these matrices do and how are they related to the eigenvalues and eigenvectors?

(i) $\begin{pmatrix} 2 & 3 \\ 1 & 4 \end{pmatrix}$ (ii) $\begin{pmatrix} 4 & -7 \\ 1 & -1 \end{pmatrix}$ (iii) $\begin{pmatrix} 3 & -1 \\ 1 & 1 \end{pmatrix}$

(iv) $\begin{pmatrix} 0 & -1 \\ 1 & 0 \end{pmatrix}$ (v) $\begin{pmatrix} -2 & 1 \\ 2 & 3 \end{pmatrix}$ (vi) $\begin{pmatrix} -1 & -2 \\ 2 & 3 \end{pmatrix}$

Visualising Matrix Transformations

Help Sheet

(1) A computer algebra system may **PLOT** vectors as if they were coordinates centred at the origin. For this work that is actually quite useful. You will need to find the equation of the line going through two vectors that are multiples of each other.

For any matrix, the vector $\begin{pmatrix} 0 \\ 0 \end{pmatrix}$ always has the result $\begin{pmatrix} 0 \\ 0 \end{pmatrix}$ so the equations are of the form $y = kx$.

A systematic approach would be to use initial vectors in a clockwise direction around the origin, determining which most nearly form a straight line with the origin. Hand drawn graphs may be most appropriate for this work.

(2) A systematic approach would be to use the unit square (i.e. the vectors $\begin{pmatrix} 0 \\ 0 \end{pmatrix}$, $\begin{pmatrix} 0 \\ 1 \end{pmatrix}$, $\begin{pmatrix} 1 \\ 1 \end{pmatrix}$, $\begin{pmatrix} 1 \\ 0 \end{pmatrix}$) and analyse the shape formed by the image vectors.

(3) If the image vector is a multiple of the original vector, this relationship must be true $\begin{pmatrix} 5 & -3 \\ 1 & 1 \end{pmatrix} \bullet \begin{pmatrix} x \\ y \end{pmatrix} = \lambda \begin{pmatrix} x \\ y \end{pmatrix}$

$\begin{pmatrix} x \\ y \end{pmatrix}$ represents the eigenvector and λ represents the eigenvalue. This equation can be **SOLVED** to give λ, although some manipulation may be needed first, such as

$$\left(\begin{bmatrix} 5 & -3 \\ 1 & 1 \end{bmatrix} - \begin{bmatrix} \lambda & 0 \\ 0 & \lambda \end{bmatrix} \right) \begin{pmatrix} x \\ y \end{pmatrix} = \begin{pmatrix} 0 \\ 0 \end{pmatrix}$$

The determinant of the matrix part of this expression will need to equal zero for this to have solutions.

(4) Using the method of question (3), the number of eigenvalues can be determined. What does a matrix having only one eigenvalue represent? What does a matrix having no eigenvalues represent?

Teaching notes

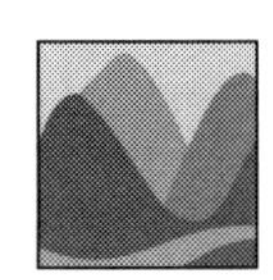

Background

This exercise is a geometrical or graphical introduction to eigenvalues and eigenvectors, rather than an algebraic or analytical approach. The aim is to give a different representation to encourage the formation of the concept of eigenvalues and eigenvectors rather than to concentrate on developing skills.

Solutions

(1) The eigenvalue is $\lambda = 2$ and the eigenvector $\underset{\sim}{e} = \begin{pmatrix} 1 \\ 1 \end{pmatrix}$ i.e. this represents the line $y = x$.

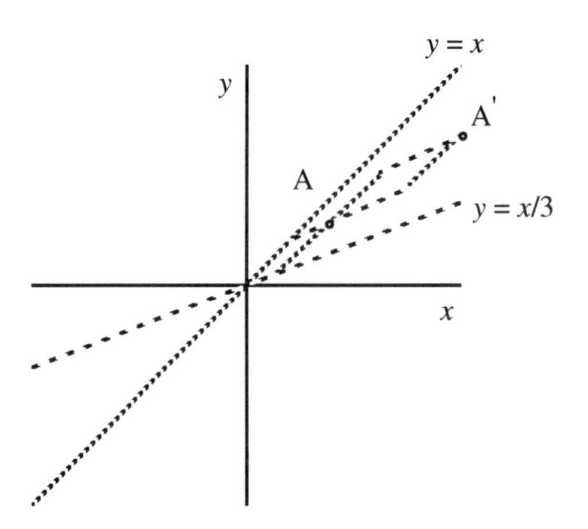

(2) The transformation has the effect of distorting the parallelogram formed from the two eigenvector lines. The distance of A from $y = x$, parallel to $y = \frac{x}{3}$, is multiplied by 4. The distance of A from $y = \frac{x}{3}$, parallel to $y = x$, is multiplied by 2. These define the position of A´.

(3) The equation to be solved is given on the help sheet. The determinant of the matrix is calculated and the resulting expression solved to give the eigenvalues above. These are then substituted into the matrix expression and manually manipulated in order to give an equation (e.g. $y = x$) for each eigenvector.

(4) (i) $\lambda = 1$ $\quad y = \frac{-x}{3}$ This represents a two-way stretch.

$\lambda = 5$ $\quad y = x$

(ii)	$\lambda = \frac{3}{2} - \frac{\sqrt{3}}{2}i$ $\lambda = \frac{3}{2} + \frac{\sqrt{3}}{2}i$		This represents a rotation, so there are no real eigenvectors.
(iii)	$\lambda = 2$	$y = x$	All points converge to a single line.
(iv)	$\lambda = i$ $\lambda = -i$		This gives a similar answer to (ii).
(v)	$\lambda = 0$ $\lambda = -4$	$y = 2x$ $y = -2x$	This gives a similar answer to (i).
(vi)	$\lambda = 1$	$y = x$	This gives a similar answer to (iii).

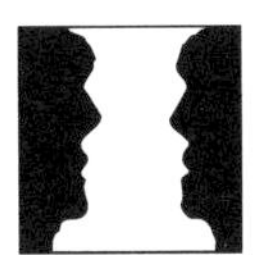

Discussion

Discussion of the importance of eigenvectors (the stretch directions) and eigenvalues (the stretch factors) should take place in relation to general points i.e. points that are not on either eigenvector.

The significance of the number of eigenvectors in relation to the geometry of the transformation could be explored.

Potential Student Difficulties

Difficulties could arise in finding eigenvectors by trial and error.

The equation that needs to be solved in order to find the eigenvalues is

$$\begin{pmatrix} 5-\lambda & -3 \\ 1 & 1-\lambda \end{pmatrix}\begin{pmatrix} x \\ y \end{pmatrix} = \begin{pmatrix} 0 \\ 0 \end{pmatrix}$$

The change in form to the following could present some conceptual difficulties

$$\det\begin{pmatrix} 5-\lambda & -3 \\ 1 & 1-\lambda \end{pmatrix} = 0$$

This may be explained by considering that the first equation has all values of x and y mapped to zero, that is that the matrix contracts all points to the origin. The area scale factor is therefore zero and hence the determinant of the matrix is also zero.

The idea of complex eigenvalues may be linked to the characteristic quadratic equation.

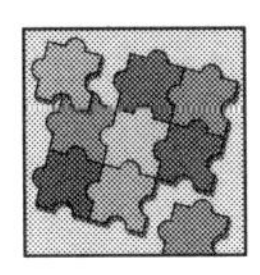

Extensions

The main area of extension would be to three dimensions. The computer algebra system will handle the algebraic manipulations, and most will give a graphical representation. In this case, however, the graphical understanding can often be best achieved away from the computer.

Blood Groups

Every person's blood can be classified into one of four types, A, B, AB and O. Typical figures, based on real data collected in the UK, suggest that people with blood group A have probability:–

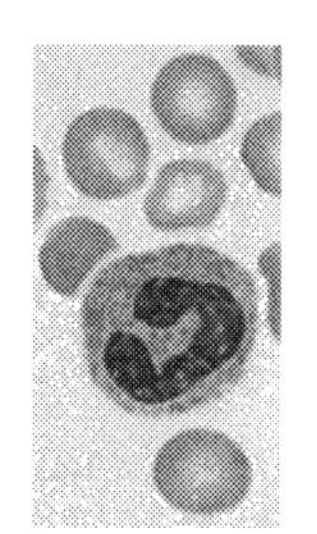

0.643 of having a child with blood group **A**.
0.030 of having a child with blood group **B**.
0.040 of having a child with blood group **AB**.
0.287 of having a child with blood group **O**.

Similar probabilities can be calculated for the other blood groups. These probabilities can be summarised in a matrix.

		One parent's group			
		A	**B**	**AB**	**O**
Offspring's group	**A**	0.643	0.119	0.465	0.250
	B	0.030	0.427	0.375	0.070
	AB	0.040	0.131	0.160	0
	O	0.287	0.323	0	0.680

(1) What is the probability that a parent with **AB** blood has a child with blood group **B**?

(2) The proportions of people of blood groups **A**, **B**, **AB** and **O** can be represented as a vector **V**. For a group of Eastern Europeans, this could be $\mathbf{V} = \begin{pmatrix} 0.450 \\ 0.130 \\ 0.060 \\ 0.360 \end{pmatrix}$.

Calculate **MV**, to find the proportions of each blood group after one generation. Calculate the proportions for 2, 3, 4... generations. Describe what happens to the vectors $\mathbf{MV}$, $\mathbf{M^2V}$, $\mathbf{M^3V}$...

(3) What happens when starting with $\mathbf{V} = \begin{pmatrix} 0.039 \\ 0.039 \\ 0.001 \\ 0.921 \end{pmatrix}$ or $\mathbf{V} = \begin{pmatrix} 0.599 \\ 0.058 \\ 0.040 \\ 0.303 \end{pmatrix}$?

(The first vector could represent indigenous South Americans, the second could represent Aborigines.)

(4) For the vectors in (2) and (3) above, find the long term proportions of each blood group.

(5) What happens to the series of matrices $\mathbf{M}$, $\mathbf{M^2}$, $\mathbf{M^3}$...? How does this relate to the answer to question 2?

Extension

(6) Try other matrices **M** – what happens?

(7) The series $\mathbf{V}$, $\mathbf{MV}$, $\mathbf{M^2V}$, $\mathbf{M^3V}$... forms a model of real population genetics. What are the assumptions that have been made?

Blood Groups

Everyone has two genes that determine their blood group (other pairs of genes determine other characteristics). A child will inherit one gene from each parent, and these determine the child's blood group.

The transition matrix defines what will happen to the blood group of the child, once the blood group of one of the parents is known. If the parent has blood group **A**, their offspring must have one of the blood groups **A**, **B**, **AB** or **O**, so these probabilities add up to 1. The probabilities depend on the population in question.

(2) Use the computer algebra system to define a 4 by 4 matrix and enter the elements horizontally. The vector **V** may be entered as a 4 by 1 matrix. It is probably easier to create these as separate statements.

Depending on the computer algebra system, it may be possible to create an expression $\mathbf{M}^n\mathbf{V}$ and then **SUBSTITUTE** different values of n.

(3) Calculating $\mathbf{M}^n\mathbf{V}$ for values of n up to 10 should give enough information. What would happen for larger values of n?

(4) The vector **V** gives the proportions of each blood group within the parent population. For any one parent, the matrix **M** gives the probabilities that offspring will have any blood group. What would **MV** represent? What effect does multiplying **M** by **MV** have?

(5) When the powers of **M** are calculated, compare the columns of each matrix. What happens to the columns? What effect does this have on different starting vectors **V**?

Extension

(6) Other transition matrices can be tried, although each column must add up to 1.

(7) **M** represents the original population and **V** represents different blood group proportions from a different population. The result of the product **MV** gives the proportion of offspring with the given blood groups when each member of the 'new' population mates with someone from the original population.

Teaching notes

Background

Many of our characteristics are inherited genetically. One of these is our blood group. We inherit two alleles, one from each parent, that determine our blood group.

These alleles come in three types, A, B and O, with proportions in a population of p, q and r. The theoretical background to the transition matrix follows the Hardy-Weinberg Law (Hardy being the mathematician G. H. Hardy), which was discovered independently by both in 1908.

Blood group **A** can have alleles AA or AO, with proportion $p^2 + 2pr$ (This assumes that the alleles are independent). Similarly, group **B** has proportion $p^2 + 2pr$; **AB** has proportion $2pq$ and **O** has proportion r^2.

If a person with alleles AB mates with a person with alleles AO, for example, there are four possible allele pairs for any offspring; AA, AB, AO, BO. The proportion for each of these offspring is p^2qr. These results give the following matrix of proportions. This matrix can be used to calculate the transition in proportions from one generation to the next.

One parent's group

$$\text{Offsping's group}\quad \begin{array}{c} \\ \mathbf{A} \\ \mathbf{B} \\ \mathbf{AB} \\ \mathbf{O} \end{array} \begin{array}{c} \begin{array}{cccc} \mathbf{A} & \mathbf{B} & \mathbf{AB} & \mathbf{O} \end{array} \\ \begin{pmatrix} p^3 + 3p^2r + pr^2 & pqr & p^2q + pqr & pr^2 \\ pqr & q^3 + 3q^2r + qr^2 & pq^2 + pqr & qr^2 \\ p^2q + pqr & pq^2 + pqr & pq(p+q) & 0 \\ pr^2 & qr^2 & 0 & r^3 \end{pmatrix} \end{array}$$

As a transition matrix is required, each column needs to be divided by the column sum (this is simplified using $p + q + r = 1$).

$$\begin{pmatrix} \dfrac{p^2 + 3pr + r^2}{p + 2r} & \dfrac{pr}{q + 2r} & \dfrac{p + r}{2} & p \\[2ex] \dfrac{qp}{p + 2r} & \dfrac{q^2 + 3qr + r^2}{q + 2r} & \dfrac{q + r}{2} & q \\[2ex] \dfrac{q(q + r)}{p + 2r} & \dfrac{p(q + r)}{q + 2r} & \dfrac{p + q}{2} & 0 \\[2ex] \dfrac{r^2}{p + 2r} & \dfrac{r^2}{q + 2r} & 0 & r \end{pmatrix} \qquad (1)$$

With values of p, q and r chosen so that they sum to 1, every column of powers of this transition matrix will tend to

$$\begin{pmatrix} p^2 + 2pr \\ q^2 + 2qr \\ 2pq \\ r^2 \end{pmatrix} \qquad (2)$$

Any such set of p, q and r will form a self-sustaining population, retaining these proportions from generation to generation. In reality, there is 'genetic drift' due to mutations or breeding with another group. The proportions of each blood group can vary quite markedly across the globe. For the indigenous British population, $p \approx 0.25$, $q \approx 0.07$ and $r \approx 0.68$.

Solutions

(1) Eastern European populations are represented by $p \approx 0.30$, $q \approx 0.10$ and $r \approx 0.60$. $\mathbf{MV} = \begin{pmatrix} 0.423 \\ 0.117 \\ 0.045 \\ 0.416 \end{pmatrix}$ $\mathbf{M}^2\mathbf{V} = \begin{pmatrix} 0.410 \\ 0.108 \\ 0.039 \\ 0.442 \end{pmatrix}$.

The series converges to $\begin{pmatrix} 0.403 \\ 0.100 \\ 0.035 \\ 0.462 \end{pmatrix}$ (3dp).

(2) For the South American group, $\mathbf{MV} = \begin{pmatrix} 0.260 \\ 0.083 \\ 0.007 \\ 0.650 \end{pmatrix}$

$\mathbf{M}^2\mathbf{V} = \begin{pmatrix} 0.343 \\ 0.091 \\ 0.022 \\ 0.543 \end{pmatrix}$ $\mathbf{M}^n\mathbf{V} \to \begin{pmatrix} 0.403 \\ 0.100 \\ 0.035 \\ 0.462 \end{pmatrix}$

For the Aboriginal group, $\mathbf{MV} = \begin{pmatrix} 0.486 \\ 0.079 \\ 0.038 \\ 0.397 \end{pmatrix}$ $\mathbf{M}^2\mathbf{V} = \begin{pmatrix} 0.439 \\ 0.090 \\ 0.036 \\ 0.435 \end{pmatrix}$

$\mathbf{M}^n\mathbf{V} = \begin{pmatrix} 0.403 \\ 0.100 \\ 0.035 \\ 0.462 \end{pmatrix}$

(3) The vectors will still converge to the same result as in question 1. This is because the series of vectors represents the blood group mix of each successive generation under certain assumptions. These mean that the 'new' group's characteristics mix with, and are gradually diluted by, the host population until the new group takes on the host's characteristics.

(4) The series of matrices also tends to values that represent the host population. Each column has the same set of values,

namely $\begin{matrix} 0.403 \\ 0.100 \\ 0.035 \\ 0.462 \end{matrix}$

Each column has the same values because, after a large number of generations, it does not matter which blood group the ancesters had. Regardless of the original mix of blood groups the final vector will always be as above, because each column of the matrix has the same form.

(5) If a matrix **M** is formed that is based on the expressions in (1), using particular values of p, q and r, the results will always reflect the columns given in (2).

However, if the matrix **M** takes any values (as long as the column sum is 1), the powers of the matrix will still tend to a limiting form that represents a viable population. The reason for this is that the original matrix would represent an unbalanced population which gradually mixes and settles down over time.

An interesting discussion could take place here. Some blood groups have *less* resistance to locally occuring diseases. How does this affect the transition probabilities for the adult offspring, as some may have succumbed to the disease in question?

(6) The assumptions are due mainly to the way in which populations would inter-breed. The model assumes that each generation of the incoming population mates exclusively with the host population. This is usually not the case, although the effect of this assumption is to speed up the change in genetic characteristics.

There is also the assumption that the new population is small, because the host population's characteristics do not change. In reality, this assumption is probably fairly valid for a British host population.

There are also genetic assumptions, for example there are no random genetic mutations.

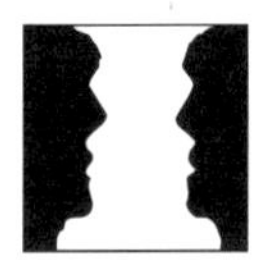

Discussion

There are several points of discussion that would help students understand the practical use of a matrix in this situation.

'What does **MV** represent?' could be asked before the help sheet was given out.

'Why can't an **AB** parent have an **O** child?' would bring out some of the underlying genetics.

The assumptions that underly the mathematical model could be discussed.

A more advanced discussion could look at the algebra behind the matrices given earlier. For example, why does a group **B** parent have a probability of *pqr* of having a group **A** child?

The parent could havc alleles *BB* or *BO*. Only the pair *BO* could give a child with blood group **A**, by mating with someone with alleles *AA*, *AO*, *AB*. These have probabilities:

$$BO \times AA \quad qr \times p^2 = p^2qr$$

$$BO \times AO \quad qr \times pr = pqr^2$$

$$BO \times AB \quad qr \times pq = pq^2r$$

The right hand expressions sum to $pqr(p + q + r) = pqr$

Misconceptions

The distinction between $\mathbf{M}^n\mathbf{V}$ and $\mathbf{M}^n$ may be difficult. These are related forms but have different perspectives within the question.

The possibility of different proportions of blood groups being sustainable (although having different matrices) may be difficult.

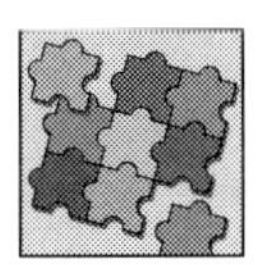

Extensions

A model of genetic characteristics could be set up from scratch. For example, the proportion of students who can roll their tongue can be ascertained experimentally. In this case, there are two alleles, the dominant tongue rolling gene *R* and the recessive gene *r*. You can roll your tongue if you have *RR* or *Rr*. You cannot roll your tongue if you have *rr*.

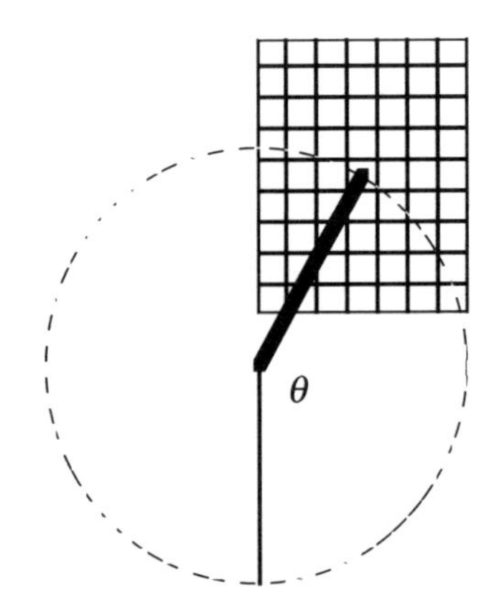

A fairground ride consists of a cage that rotates around the central pivot. The cage is propelled by the people inside redistributing their weight, just as in an ordinary swing. However, the cage is free to move completely around a circular orbit.

The radius of the arm is 1.5m and the mass of the cage and occupants is about 400 kg.

(1) The velocity at the bottom of the circle is $u = 5\text{ms}^{-1}$. What is the kinetic energy at that point? What would the potential energy be at the top of the circle? Will the cage be able to go over the top in this situation? How does this depend on the value of u?

(2) Now model the motion in a more general way. What is the velocity at an angle θ to the downward vertical? Try using different values of u.

(3) What is the acceleration of the cage at an angle θ to the downward vertical?

(4) Use your results to write a report on the motion of the cage. Include equations and graphs to describe the motion. Try to include both velocity and acceleration in your report.

(1) To decide on the potential energy at the top, you need to decide where the datum for the potential energy will be (this is the height of the line where you take potential energy as being zero). If you use conservation of energy, the total energy at each point needs to be evaluated.

(2) In order to form an equation for v, the velocity at an angle θ to the downward vertical, you need expressions for the potential and kinetic energy at the bottom and at a general point. The computer algebra system can **SOLVE** these, once equated, to give an expression for v in terms of θ and the other parameters.

(3) To find the acceleration it is probably easiest to work in angular velocity, using $\frac{d\theta}{dt} = \frac{v}{r}$ for the angular velocity.

Use the computer algebra system to **SUBSTITUTE** expressions using $\frac{d\theta}{dt}$ so that the expressions have constants θ and $\frac{d\theta}{dt}$ in them.

$-r\left(\frac{d\theta}{dt}\right)^2$ (the radial acceleration) can be found using the computer algebra system, probably by **SUBSTITUTING**.

$r\frac{d^2\theta}{dt^2}$ (the tangential acceleration) can be found by

DIFFERENTIATING the expression found in (2) and then **SUBSTITUTING** an expression for $\frac{d\theta}{dt}$ so that only θ and constants remain.

(4) Your report should include assumptions, a detailed analysis and practical considerations, for example the acceptable value for acceleration for such a fairground ride before it becomes potentially dangerous.

Teaching notes

Background

The aim of this activity is to use a computer algebra system to set up circular motion equations and analyse the motion using graphs and calculus.

For a vertical circle, conservation of energy is the key strategy. This can be calculated in polar coordinates.

$$\text{KE} = \frac{1}{2}mr\left(\frac{d\theta}{dt}\right)^2$$

PE = $mgr(1 - \cos(\theta))$ measured with the datum at the bottom of the circular path.

A computer algebra system will need θ to be defined as a function of time t.

Solutions

(1) The kinetic energy is $\frac{25}{2}m$ N

The potential energy is $3mg$ N, taking the potential energy to be zero at the bottom of the circle.

Thus the cage will not make a complete orbit, as 12.5 is less than $3g$. The cage will make a complete orbit if $u^2 > 6g$, i.e. $u > 7.7 \text{ ms}^{-1}$ approximately.

(2) Assuming that the cage has a velocity u at the bottom of the circle and that friction is negligible, the following results are obtained.

Energy at $\theta = 180°$ $\quad \frac{1}{2}mu^2$

Energy at $\theta = 0°$ $\quad \frac{1}{2}mv^2 + \frac{3}{2}mg\,(1 - \cos\,(\theta))$

These expressions can be **INPUT** into the computer algebra system and **SOLVED** for v.

$$v = \pm\sqrt{3g\cos\,(\theta) - 3g + u^2}$$

Given the value of u, a graph of v against θ can be produced. A family of such curves will probably be easier for students to understand than a three dimensional graph.

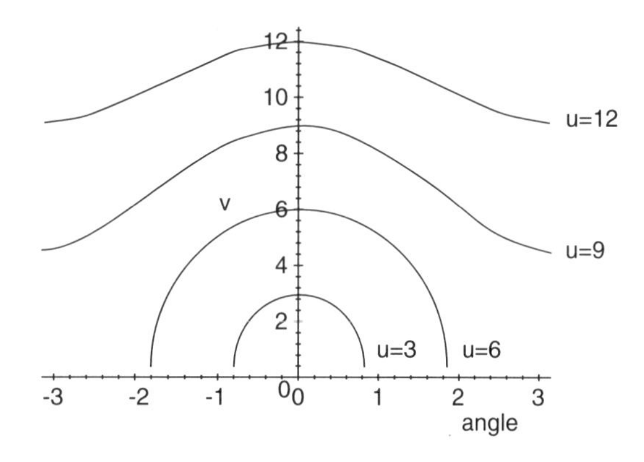

Here, v positive has been plotted. The negative will give the reflection of this curve in the θ axis. The four curves are for u = 3, 6, 9 and 12 ms^{-1} and show the qualitatively different behaviour (complete revolutions or not) depending on the value of θ.

(3) Using $v = r\omega$, the equation for v can be rewritten as (without the $\pm$ sign – see below)

$$r\frac{d}{dt}\theta = \sqrt{3g\cos(\theta) - 3g + u^2} \qquad (1)$$

Where θ has been defined as a function of t in the computer algebra system. Note that r has been kept as the general letter on the left hand side, to emphasise the links with the expressions for acceleration.

To find the radial acceleration, $-r\left(\frac{d}{dt}\theta\right)^2$, (1) is squared and divided by r to give $-\frac{3g\cos(\theta) - 3g + u^2}{1.5}$

To find the tangential acceleration, $r\left(\left[\frac{d}{dt}\right]^2\theta\right)$, expression (1) is differentiated and simplified as necessary to give

$$\frac{3g\left(\frac{d}{dt}\theta\right)\sin(\theta)}{2\sqrt{3g\cos(\theta) - 3g + u^2}}$$

An expression for $\frac{d}{dt}\theta$ can be found using (1) and substituted in the expression above (the $\pm$ signs have cancelled here) to give

tangential acceleration = $-g\sin(\theta)$

The expressions can then be used as the basis for a report of the motion, picking out the salient features of velocity and acceleration. Graphs of the acceleration expressions can be used to support this, plotted for various values of u.

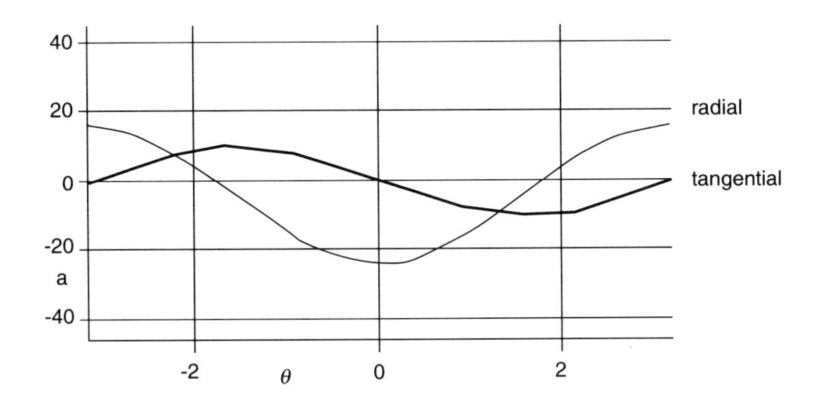

The above graph shows the components of acceleration for $u = 6$, and the graph below shows the components of acceleration for $u = 9$. In each case the narrow line represents the radial acceleration and the broad one represents tangential acceleration.

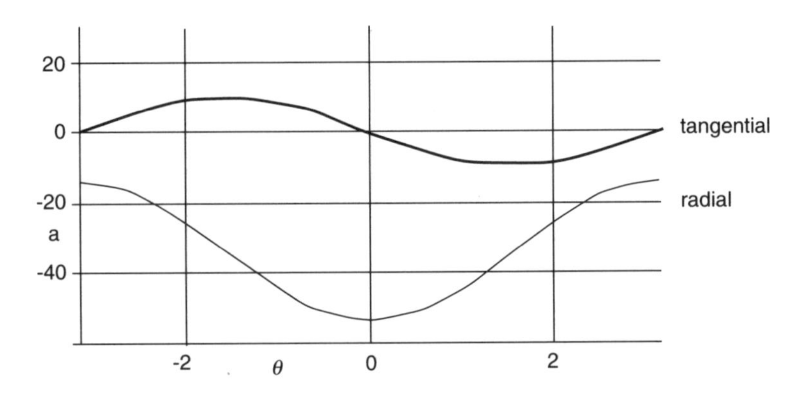

Note that neither graph has meaning when the radial acceleration is *positive*, as the cage does not reach this part of the arc.

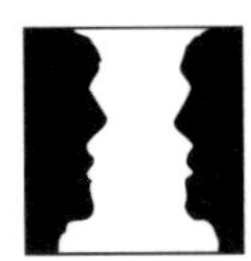

Discussion

The coordinate system to use could be discussed. Initially, rectangular coordinates are easier, but the switch to polar coordinates is needed if acceleration is to be readily calculated.

The assumptions that are needed for the model could also be discussed and justified.

The need for a potential energy datum, and the sensible choice of its position may need to be addressed.

Computer algebra systems may not be able to cope with ± cases, for example when square rooting, except by listing both cases seperately. In these situations, the entire analysis can be performed for both cases, but discussion of ways to avoid this duplication, for example by consideration of symmetry, would be worthwhile.

The important features of the motion, such as velocity being zero or a component of acceleration being zero, may need to be emphasised to the students.

More advanced students may like to consider the derivation of the acceleration formulae fromfirst principles. This will involve the derivatives with respect to time of the base vectors $\hat{\underline{r}}$ and $\hat{\underline{\theta}}$, namely these should be $\dot{\theta}\hat{\underline{\theta}}$ and $-\dot{\theta}\hat{\underline{r}}$.

Potential Student Difficulties

The use of polar coordinates may be unfamiliar and need further development. Practical experiments on the effect of centripetal acceleration could be performed using a turntable.

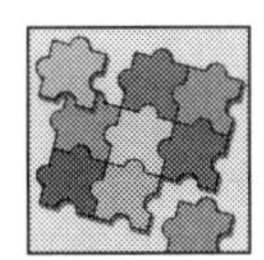

Extensions

Extensions could be theoretical, where students are asked to derive the components of the angular acceleration from the base vectors, as discussed above.

An extension to the model could be to consider the motion of the centre of gravity of the cage. This must be a distance below the pivot so that the cage remains upright – perhaps 30 or 50cm. This will alter the kinetic and potential energy values and give a different description of the motion.

Swing Safety

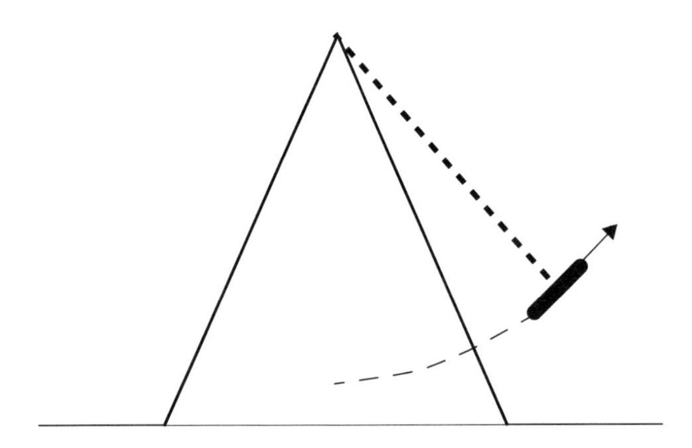

Children's play areas should have a safety surface. This is a softer surface extending round the play equipment designed to reduce the severity of injury should a fall occur. The safety area needs to be large enough to ensure that a falling child lands on it, but the cost means that it should be kept as small as possible. How large should the safety area round a swing be?

(1) Decide on the constants and variables that would apply to a child using a swing. Explain the assumptions that you need to make and justify them.

(2) Set up a mathematical model that allows you to work out the distance that a falling child would cover.

(3) Analyse your model and write a report to summarise your results. You should consider the limitations in your model and ways in which you could refine it.

Extension

(4) A swing might already have a safety surface around it. However, this might too small. How could the swing be altered (easily and cheaply) so that the safety surface will now be adequate?

(1) The most important (and probably least justifiable) assumption should be that the child on the swing can be modelled as a single particle at, or just above the swing's seat.

Variables should include θ for the angle that the swing has reached when the fall takes place. Some value for the total energy of the swing must be used as well – a small amplitude swing will not project a child far! For example, you could use θ_{max}, the maximum amplitude, or E, the total energy.

(2) The mathematical model falls into two sections. The first is to model the swing using circular motion. This will allow you to work out the horizontal and vertical values of the velocity at the instant the child falls. Using a computer algebra system, **INPUT** expressions for the potential and kinetic energies at a reference point and at the point where the child falls. Use the computer algebra system to combine these into a single equation and **SOLVE** to find the velocity.

The second part is to model the child as a projectile. From the first part, you know the velocity and position at the beginning of the free fall. **INPUT** expressions that give the horizontal and vertical positions at time t. The vertical position expression can then be **SOLVED** to give the time of flight. This value of t can be **SUBSTITUTED** back to give the range.

(3) It would be helpful to **PLOT** a graph of range against θ, the angle of fall. What angle of fall gives the largest range? Is this always the case for other values of θ_{max}?

DIFFERENTIATION can be used to find the maximum range for a particular value of θ_{max}.

Extension

(4) Extend the problem by considering the use of parameters rather than actual values. For example you could use c for the length of the swing's chain, in order to find the effect that c has on the expression for the range.

Teaching notes

Background

This activity can be used as a piece of A level coursework. The original idea formed the assignment for *Modelling with Force and Motion*, an SMP 16–19 text.

Solutions

The solutions given here are at their most general level, where all parameters are unknown rather than constants. Students should make many assumptions that will make the problem easier. Many students may study the circular part of the motion generally, but then substitute particular angles into their expressions in order to get a number of particular cases of the projectile motion. However, if the students can determine the steps that are needed in the solution, they should be able to use the computer algebra system to produce the necessary results. The resulting expressions can be daunting!

The solutions give one particular example for the graphical analysis. Other values of the parameters can be substituted readily for other scenarios. The parameters used are: chain length c m; seat height above the ground h m; $g = 10\text{ms}^{-1}$.

The angle to the downward vertical is θ and the speed of the swing at a general point is v.

Taking the potential energy datum at the bottom of the swing's arc and using the law of conservation of energy,

$$\frac{1}{2}mu^2 = \frac{1}{2}mv^2 + cmg\,(1 - \cos[\theta])$$

which can be **SOLVED** to give

$$v = \sqrt{u^2 - 2gc\,(1 - \cos[\theta])}$$

Note that only the positive velocity case is considered here, as the negative velocity represents the swing travelling in the other direction.

If the child then falls from the swing when it is at an angle θ to the downward vertical, its motion is determined by

position $\qquad \begin{pmatrix} c\sin[\theta] \\ c\,(1 - \cos[\theta]) + h \end{pmatrix}$

velocity $\qquad v\begin{pmatrix} \cos[\theta] \\ \sin[\theta] \end{pmatrix}$

and acceleration $\begin{pmatrix} 0 \\ -g \end{pmatrix}$

where the coordinate system has the x component horizontal and the y component vertical, with the origin directly under the swing when it is at rest.

Using the standard projectile motion equations, these give the position at a general time t secs as

$$\begin{pmatrix} c\sin[\theta] \\ c(1-\cos[\theta]) + h \end{pmatrix} + vt\begin{pmatrix} \cos[\theta] \\ \sin[\theta] \end{pmatrix} + \frac{t^2}{2}\begin{pmatrix} 0 \\ -g \end{pmatrix}$$

When the child hits the ground, the vertical component is zero and this component of the expression can be **SOLVED** to give the time of flight, t secs.

$$t = \frac{\sqrt{-2gc(\cos[\theta])^3 + (u^2 - 2gc)(\sin[\theta])^2 + 2(c+h)g}}{g} + \frac{\sin[\theta]\,v}{g}$$

Again, only the positive value of t is used as the other value does not represent a physically possible situation.

This value of t can the be substituted into the horizontal component in order to give the range.

$$R = c\sin[\theta] + \frac{v}{g}\left(\sqrt{-2gc(\cos[\theta])^3 + (u^2 - 2gc)(\sin[\theta])^2 + 2g(c+h)} + \sin[\theta]\,v\right)$$

where

$$v = \sqrt{u^2 - 2gc(1-\cos[\theta])}$$

The expression for R can be plotted for various values of u to give the following curves. θ is in radians .

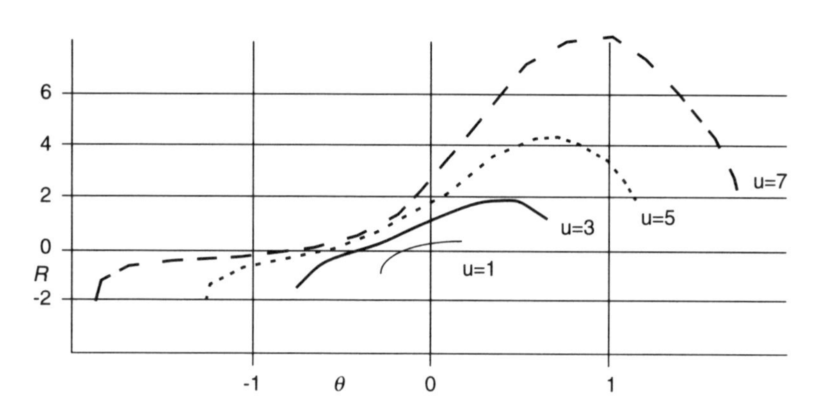

Here, c = 2m, h = 0.5m, g = 10ms^{-2} and u = 1, 3, 5 and 7ms^{-1} give the curves from bottom to top. Note that the value of u = 7ms^{-1} would have the swing reaching approximately 103° or 1.8 radians, which is probably impractical.

The cases that were not considered above, with v and t negative, will give symmetrical results.

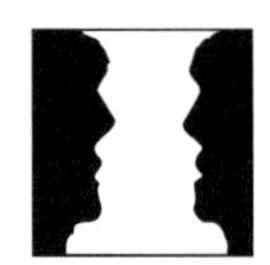

Discussion

A discussion will need to take place of the assumptions that need to be made in order for the model to be relatively simple but realistic. The choice of variable and constant parameters is important, with the emphasis on keeping as many as constant as possible. This will ease the student's interpretation of the results as well as making it possible to view the expressions in full on the screen!

The division of the motion between circular and parabolic portions can be discussed. The key aspect is the equivalence of the boundary conditions i.e. position and velocity where the type of motion changes.

The appropriate coordinate system and the most appropriate position for the axes could also be considered.

In the model used above, the position of the centre of gravity is taken to be on the seat. The necessity for this assumption and its validity could be discussed.

Potential Student Difficulties

The fact that both components of position and velocity must be equal at the boundary between the different parts of the motion may be an area that students find difficult.

The choice of coordinate systems may show up some misconceptions. For example if the coordinate system is chosen so that the swing at rest is at (0,0), then the range has to be calculated as the distance travelled to a point $(R,-0.5)$.

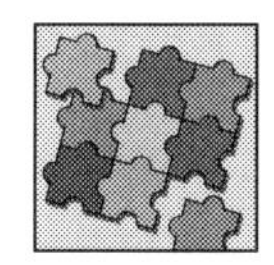

Extensions

The main form of extension would be to replace some constants by parameters in order to refine the model. This would probably be necessary for the suggested extension given on the activity sheet.

No Turning Back

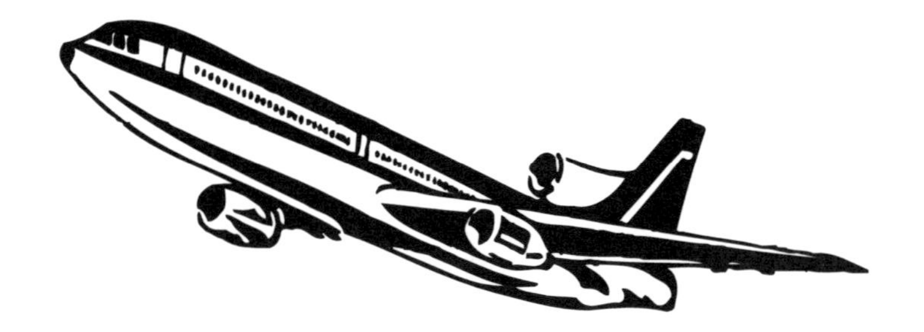

You are the pilot of an aircraft that is capable of cruising at a constant speed in still air. You have enough fuel to give four hours of flight. You need to decide on the maximum distance that you can fly from your base and still return home safely when there is a constant wind velocity. You will also need to provide details of the direction of flight.

(1) Set up a model for the aircraft's flight. You will need to consider the assumptions that you will make, the constants and the variables. Start with a very simple model, enabling you to make it more complex later.

(2) Work out the actual velocity of the aircraft taking account of the wind, on both the outward and return journeys.

(3) Using the velocities from above, calculate the distance that the plane will be able to travel safely in order to return to its base.

(4) Comment on your results and any limitations that your model has.

Extension

(5) Decide on how you can change your model to provide a more realistic analysis of the problem.

(1) When setting up the model, you may like to consider:–
What are the assumptions that you need to make?
What are the variables?
Which variables can you fix to simplify the model?

Initially, you may like to fix the plane speed, the wind speed and have the plane flying directly into the wind. This can be solved by drawing velocity diagrams and using a calculator.

The model can be refined so that the wind speed is a variable w kmh^{-1}.

(2) Using vectors, draw velocity diagrams for the outward and return legs of the journey.

(3) Using a computer algebra system, **INPUT** expressions for the distance travelled, equate these and **SOLVE** to get the time t hours as a function of w kmh^{-1}, and then get the distance d km as a function of w kmh^{-1}. **PLOT** graphs of both functions and use them to explain your answer to the problem.

A different refinement would be to call the plane's speed p kmh^{-1}, but keep the wind speed a constant. In the same way **INPUT** an equation that shows that the distance out is the same as the distance back. **SOLVE** this equation and **PLOT** graphs to help explain the results.

You might like to try having a wind speed of w kmh^{-1} and a plane speed of p kmh^{-1}.

(4) Comment on your results and any limitations of your model.

Extension

(5) The plane does not have to fly straight into the wind. It can fly at an angle to the wind. Draw careful velocity diagrams and use component velocities to **INPUT** expressions (either as vectors or separate components) for the distance travelled. As before, equate these and **SOLVE** the expression to find the time of flight and then the distance travelled.

Teaching notes

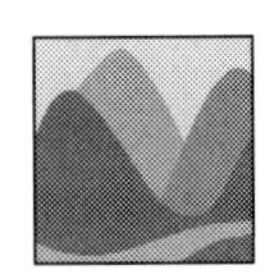

Background

This activity is based on the 'Point of No Return' problem from the SMP 16–19 textbook *Newton's Laws of Motion*. It could form an assessed piece of coursework for some courses.

Solutions

A model with the plane flying into the wind, which has velocity w kmh^{-1}.

In step (3) the time of flight outwards is t hours, the time to return is $4 - t$ hours. With a plane speed of 100kmh^{-1}, flying into the wind initially, the distance outwards is $t(100\text{-}w)$km. Similarly the distance inwards can be calculated and the two expressions equated.

$$t\,(100 - w) \;=\; (4 - t)\;(100 + w)$$

This equation can be solved easily using a computer algebra system to find t in terms of w.

$$t \;=\; \frac{(100 + w)}{50}$$

The distance flown can also be expressed in terms of w.

$$d \;=\; \frac{(100 + w)\;(100 - w)}{50}$$

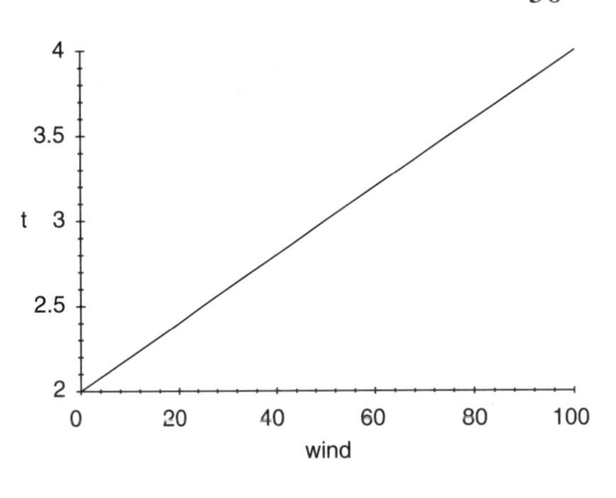

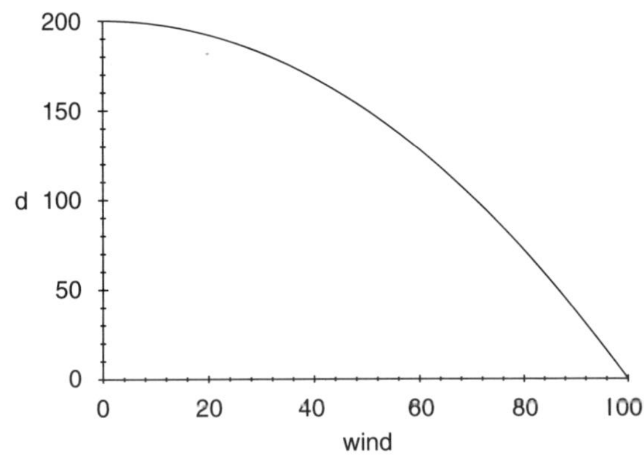

N.B. These graphs can be extended to negative wind speeds before being interpreted.

A model with the plane flying at speed p kmh^{-1} into the wind.

Step (3) can be worked in a similar way to above; the distance outwards can be expressed as $t(p - 20)$km given a wind speed of 20kmh^{-1}. The inward and outward distances are equated

$$t\,(p - 20) \;=\; (4 - t)\;(p + 20)$$

and solved to find t in terms of p.

$$t \;=\; \frac{2\,(p + 20)}{p}$$

The distance d can then also be expressed in terms of p.

$$d = \frac{2\,(p+20)\,(p-20)}{p}$$

The graphs below can also be extended to negative values of plane speed. Some care is required in the interpretation of the graphs, as in this case an absolute value of plane speed less than 20kmh^{-1} does not give a realistic solution.

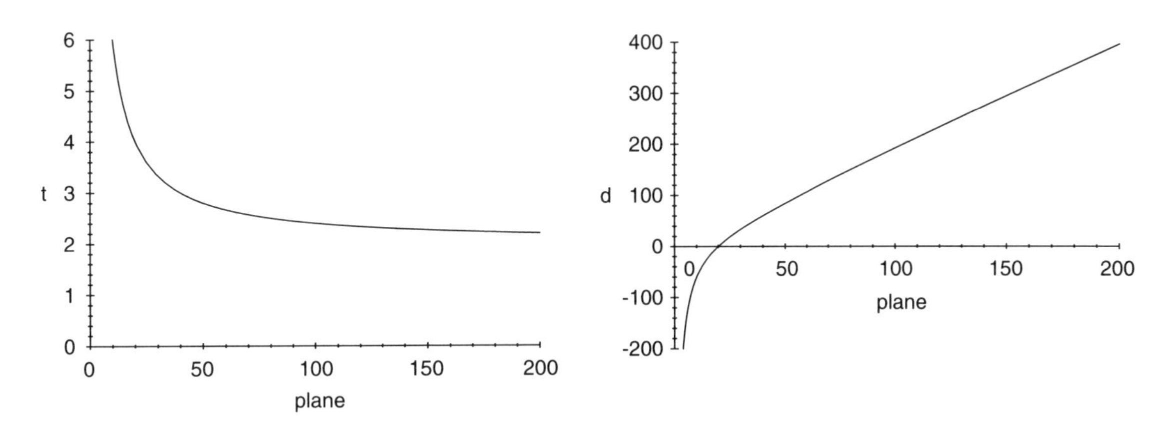

A model with both p and w variable.

Following the same routine for step (3), a computer algebra system will easily produce expressions for t and d. It is possible to draw 3-D surface plots but these are difficult to interpret, so a family of curves for various values of p and w may be a better alternative.

The equation can be solved to find t and d in terms of p and w.

$$t = \frac{2\,(p+w)}{p}$$

$$d = \frac{2\,(p+w)\,(p-w)}{p}$$

Extension – allowing the angle of flight to vary.

Careful velocity vector diagrams are needed in step (5) to correctly analyse the aircraft's route.

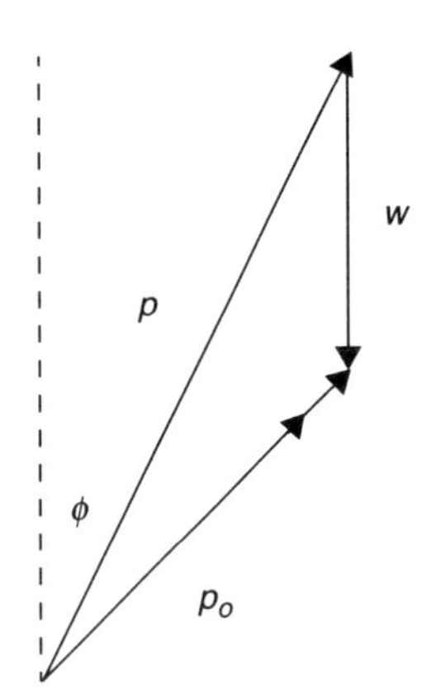

Taking **p** kmh^{-1} as the plane's velocity, **w** kmh^{-1} as the wind's velocity (always taken as a north wind), then the velocity diagram gives $\mathbf{p}_o$ kmh^{-1}, the outward velocity of the plane relative to the ground.

$$\mathbf{p}_o = \begin{pmatrix} p\sin[\phi] \\ p\cos(\phi) - w \end{pmatrix} \tag{1}$$

The return velocity of the plane, $\mathbf{p}_r$ kmh^{-1} must have the opposite direction to $\mathbf{p}_o$ although its magnitude could be different. Hence

$$\mathbf{p}_r = -\lambda\mathbf{p}_o$$

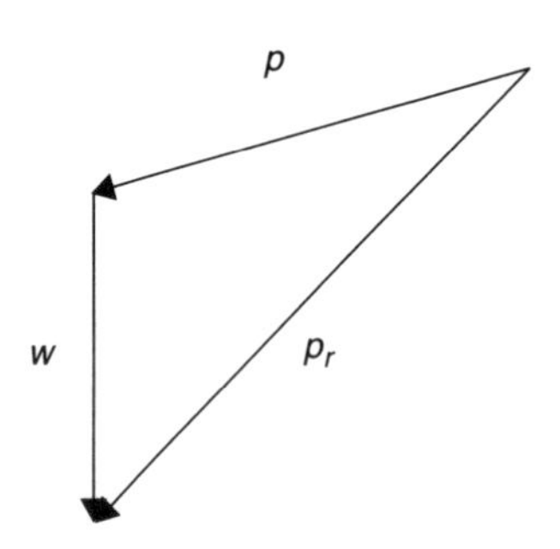

The velocity diagram shows that the plane's velocity relative to the air can be expressed as

$$\mathbf{p} = -\lambda\mathbf{p}_o - \mathbf{w}$$

which gives

$$\mathbf{p} = \begin{pmatrix} -p\lambda\sin(\phi) \\ -p\lambda\cos(\phi) + w(1+\lambda) \end{pmatrix} \tag{2}$$

The plane's airspeed is a constant, p kmh^{-1}, so the magnitude of **p** can be equated to p using a computer algebra system to give

$$(p\lambda\sin(\phi))^2 + (w(1+\lambda) - p\lambda\cos(\phi))^2 = p^2$$

The computer algebra system will solve the equation for λ with little difficulty, giving

$$\lambda^2(p^2 + w^2 - 2pw\cos(\phi)) + \lambda(2w^2 - 2pw\cos(\phi)) + w^2 - p^2 = 0$$

Given values of p kmh^{-1} and w kmh^{-1}, this quadratic can be solved to give λ in terms of ø. Using the return speed is λ times the outward speed, $t|\mathbf{p}_o| = (4-t)\lambda|\mathbf{p}_o|$ or $t = \dfrac{4\lambda}{1+\lambda}$

The distance travelled can then be calculated using (1) and the direction of flight for the return journey using (2).

Below are graphs of the time of outward flight and the maximum distance attainable for $p = 100\text{kmh}^{-1}$ and $w = 50\text{kmh}^{-1}$.

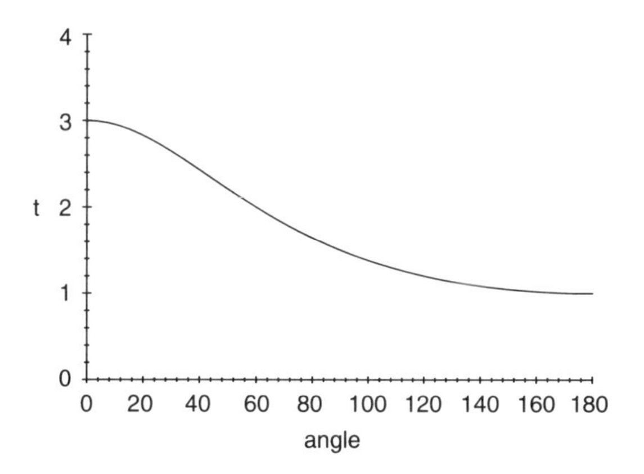

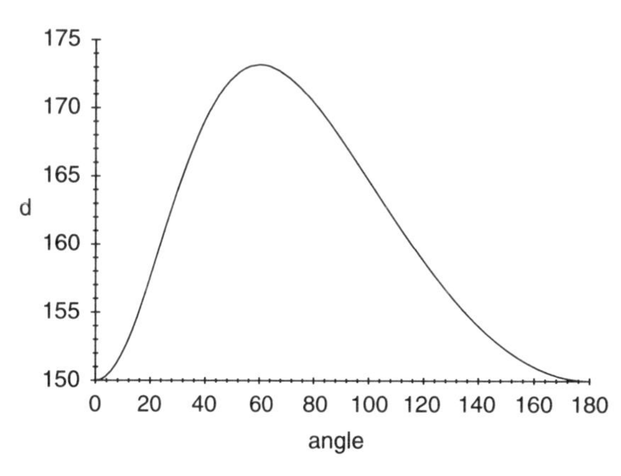

The use of a computer algebra system allows the students to continue to refine the model that they are using without becoming bogged down in difficult algebra. Once they have mastered the velocity diagrams and worked out the equations to be solved, the remaining steps do not get any more demanding as the computer algebra system can perform the necessary manipulations.

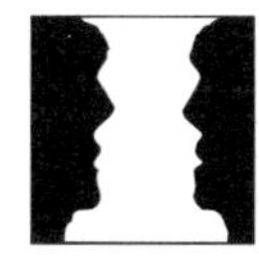

Discussion

The quantities that can be equated for the outward and return legs of the journey will probably need some exploration. Also, the assumptions that are made in order to produce a simple mathematical model will need careful justification.

Potential Student Difficulties

The combination of the velocity of the plane relative to the air with the wind speed in order to find the velocity of the plane relative to the ground may cause some difficulties. The interpretation of the graphs, particularly those parts which are not physically meaningful, may also be an area that students find hard to comprehend.

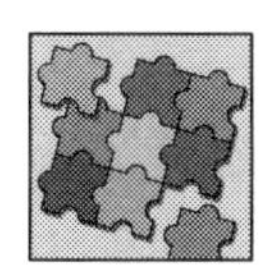

Extensions

The problem can be extended by making it more general. It may be interesting to note that this type of problem was often solved during the Second World War, but for the case where the base was a moving aircraft carrier!

Modelling the Sine Function

When calculators compute values for functions like sine, they use a polynomial which is very close to the sine function, accurate to at least 8 decimal places. Develop your own polynomial function which approximates the sine function in the domain 0 to 2π. You can do this in many ways – two are suggested here.

(1) Examine the graph of the sine function and construct a polynomial with the same roots and the same height. Remember that $f(x) = kx(x - a)(x - b)$ has roots at 0, a and b.

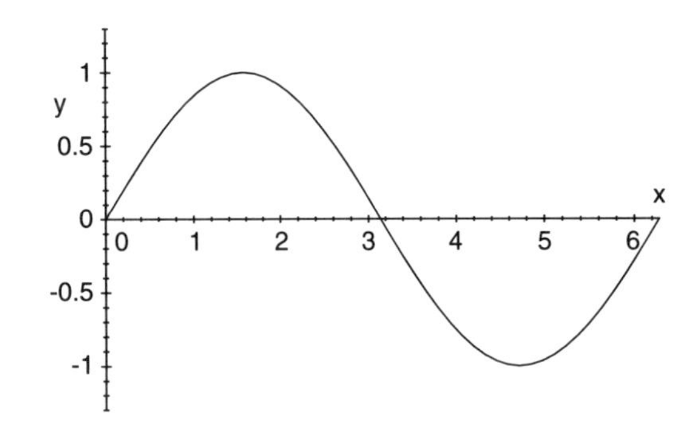

(2) Alternatively, start with a polynomial $f(x) = ax^3 + bx^2 + cx + d$, then generate a number of simultaneous equations by equating this polynomial with the sine function at particular points. For example;

$$\sin(0) = 0 \Rightarrow f(0) = 0 \Rightarrow d = 0$$

$$\sin(\pi) = 0 \Rightarrow f(\pi) = 0 \Rightarrow a\pi^3 + b\pi^2 + c\pi + d = 0$$

With four such equations, solve them simultaneously to isolate values for a, b, c and d and substitute these co-efficients into $f(x)$.

(3) Develop three approximating polynomials of different orders. You may want to use information concerning the derivatives of the sine function at particular points, as well as particular values of the function; e.g. $f'(0) = 1$, $f'\left(\frac{\pi}{2}\right) = (0)$

(4) Determine the accuracy of each polynomial by comparing the difference between each candidate's function and the sine function. You can do this graphically or numerically.

Modelling the Sine Function

(1) If you want to explore a solution graphically, then use the three roots of the sine function between 0 and 2π to construct a polynomial with the same roots. This should give you a solution that crosses the x-axis in all the right places, but probably does not have the same maximum or minimum. You would then need to multiply your function by a particular value to give it maxima and minima of 1 and –1.

Also try polynomials with double roots;. e.g. $(x-1)(x+2)$ has single roots at $x = 1$ and $x = -2$, while $(x-1)^2(x+2)^2$ has double roots at the same points.

(2) The analytic method is to define a a function as a polynomial with unknown coefficients; e.g.

mysine(x): $= ax^3 + bx^2 + cx + d$

You can then **INPUT** four equations which describe how you want to constrain your estimate function and use the computer to **SOLVE** all four equations to give you values for the four unknowns:

$$\{\text{mysine}(0) = 0, \text{mysine}(\pi) = 0, \text{mysine}(2\pi) =$$
$$= 0, \text{mysine}\left(\frac{\pi}{2}\right) = 1\}$$

SUBSTITUTING the values of a, b, c and d back into mysine(x) then gives you an approximate sine function. Try polynomials of higher order using a similar method.

To determine models using polynomials of a higher order, use more information concerning the sine function to generate further equations, then use the CAS to **SOLVE** this larger set of simultaneous equations.

(4) To get a rough idea of the accuracy of your estimate functions, **PLOT** these with sin(x). Another way is to **PLOT** the difference between each of your solutions and the sine function (e.g. **PLOT** mysine(x) – sin(x)) and examine these between 0 and 2π to see how they vary.

A more accurate way to measure 'closeness' is to examine the difference between the sine function and an estimate function more analytically. You could **DIFFERENTIATE** this difference function and isolate the maxima and minima to find the limits of accuracy between the two functions.

Teaching notes

Background

The trigonometric function is often shrouded in numerical mystery as students always have to rely on a calculator to generate values. Even if they study the series functions – via Taylor expansion or otherwise – this only goes some way to turn trigonometric function into something 'calculatable'. In this activity, students are asked to approximate the sine function over a fixed domain, using polynomial functions. Developing polynomial approximations should give students a deeper understanding of how polynomial and transcendental functions are related over a fixed domain.

Solutions

(1–2) Either method should give the following cubic solution if the sine function is sampled at the roots and is given a maximum of 1. If the analytic method is used, four simultaneous equations are required; e.g. $f(0) = 0$, $f(\pi) = 0$, $f(2\pi) = 0$ & $f(\pi/2) = 1$. Adding the further constraint of $f(3\pi/2) = -1$ to a quartic produces in the same cubic; i.e. the cubic already has the point $(3\pi/2,-1)$

$$\frac{16x}{3\pi} - \frac{8x^2}{\pi^2} + \frac{8x^3}{3\pi^3} = \frac{8}{3\pi^3}x(x-\pi)(x-2\pi)$$

If $f'(\pi/2) = 0$ is imposed on a quintic, together with the five constraints above, this generates the polynomial below. Adding the further constraint of $f'(3\pi/2) = 0$ to a polynomial order 6 produces in the same quintic; i.e. the quintic already has the derivative $f'(3\pi/2) = 0$:

$$\frac{8x}{3\pi} + \frac{28x^2}{9\pi^2} - \frac{116x^3}{9\pi^3} + \frac{80x^4}{9\pi^4} - \frac{16x^5}{9\pi^5}$$

Students may develop different solutions if they impose different constraints on their function.

(3) Examination of the graph of the difference function is a good way to explore the accuracy of students' model sine functions. Students can zoom in on maxima and minima to find out the limits of accuracy. Graphs of the difference functions for four different models are shown below – together with their limits of accuracy:

i) Difference function of 'sin(x) – cubic' gives; minimum point (0.590, –0.181) and maximum point (5.693, 0.181)

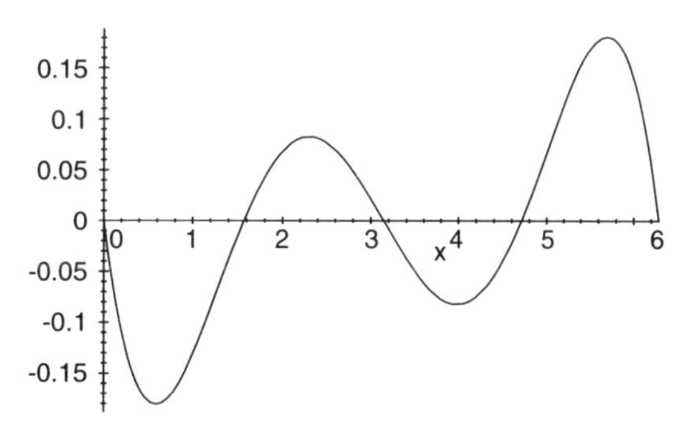

ii) Difference function of 'sin(x) – quintic' gives; minimum point (5.899, –0.0238) and maximum (0.385,0.0238)

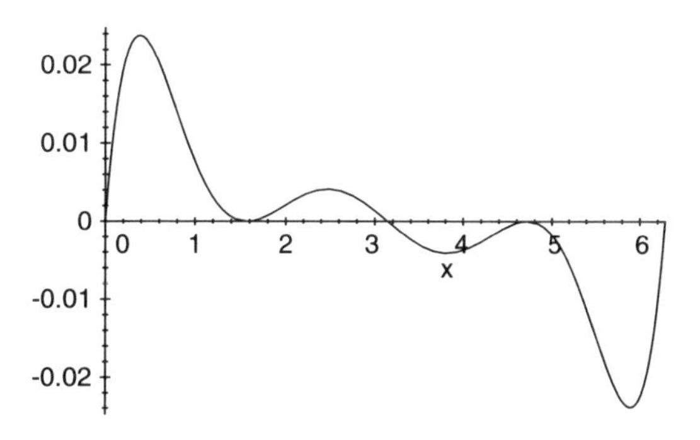

iii) Using a polynomial of order 7 with the following constraints:
$f(0) = f(\pi) = f(2\pi) = 0, f(\pi/2) = 1, f(3\pi/2) = -1, f'(\pi/2) = f'(3\pi/2) = 0, f'(0) = 1$
The difference function gives;
minimum point (0.611, 0.000974) and
maximum (5.672, –0.000974)

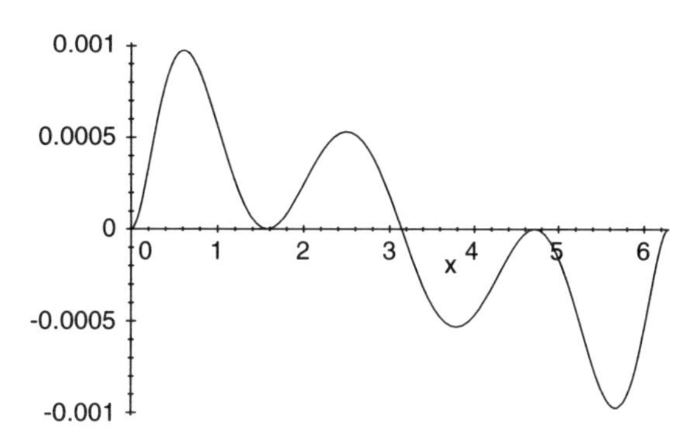

iv) Using a polynomial of order 9 with the following constraints:
$f(0) = f(\pi) = f(2\pi) = 0, f(\pi/2) = 1, f(3\pi/2) = -1,$
$f'(\pi/2) = f'(3\pi/2) = 0, f'(0) = 1, f'(\pi) = -1, f'(2\pi) = 1$
The difference function gives;
minimum point (0.521, –0.0000642) and maximum point (5.762, 0.0000642)

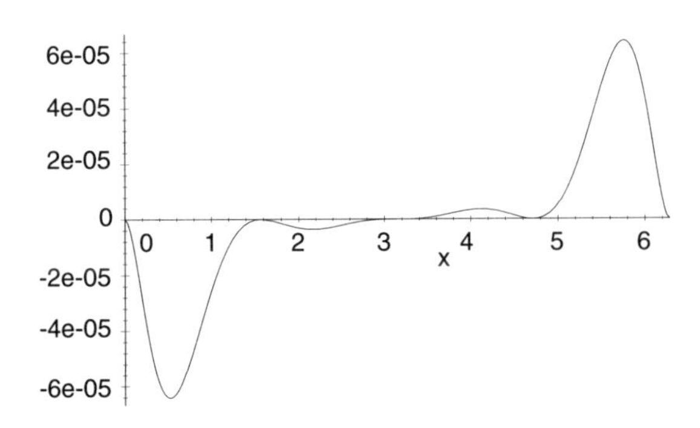

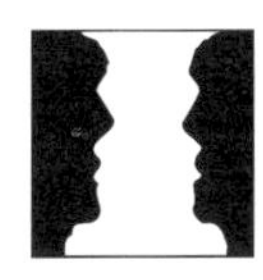

Discussion

Clearly, the estimate function only models the sine function over a fixed domain – and this is an important limitation to bring out. Students can use their estimate function, $f(x)$ say, to define a periodic function over the real numbers by defining a function as:

mysine (x) : = f(x mod 2π)

Activity Worksheet

Solving Equations with Tangents

If we cannot solve an equation analytically (i.e. by an exact formula), we need to use numerical methods. Using tangents can be an efficient way of solving equations. For example if we wish to solve the equation $x\sin x = 1$, that is the same as solving $x\sin x - 1 = 0$.

(1) To help find where $x\sin x - 1 = 0$, a plot of $x\sin x - 1$ reveals **a** solution in the interval [1,2], there are of course other solutions but we will concentrate on this one. **PLOT** this on your CAS.

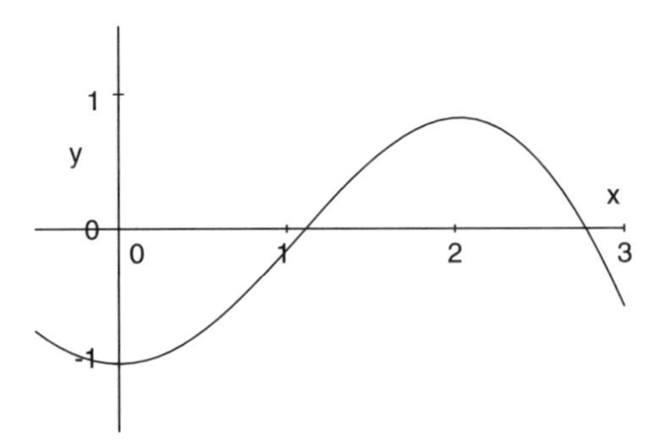

(2) We will guess that $x = 1$ is an approximate solution, but is there a better approximation? Find the equation of the tangent at the point $x = 1$ (using the techniques found in the Equation of a Tangent worksheet).

(3) Plot the equation of the tangent, visually check that it is in fact the tangent at the point $x = 1$.

(4) Solve the equation of the tangent, where $y = 0$. The value of x should be a better approximation to the solution.

(5) Use this new value of x to construct a new tangent, repeat (3) and (4) with this new value.

(6) Keep on repeating this process until there is no change in the solution to 6 decimal places.

(7) Solve the following equations using tangents, with the given first guess.

(a) $e^{\sin x} + x = 0, x = 1$ (b) $2^{\sqrt{x}} \ln x - \sin x = 0, x = 1$

Extension

(8) Use the equation in 7(a), but with a starting point $x = 2$ to investigate when this method can fall down. Write a report on when this method is inefficient or even fails!

Solving Equations with Tangents

Help Sheet

When solving an equation of the form $f(x) = 0$, if you plot the graph of the expression, finding where it is zero is the same as seeing where the graph cuts the x axis. This method needs a starting point, a good guess at the solution. The better the guess the more efficient the method! The following picture describes the process, use an initial guess and draw the tangent to the curve at the point corresponding to this initial guess.

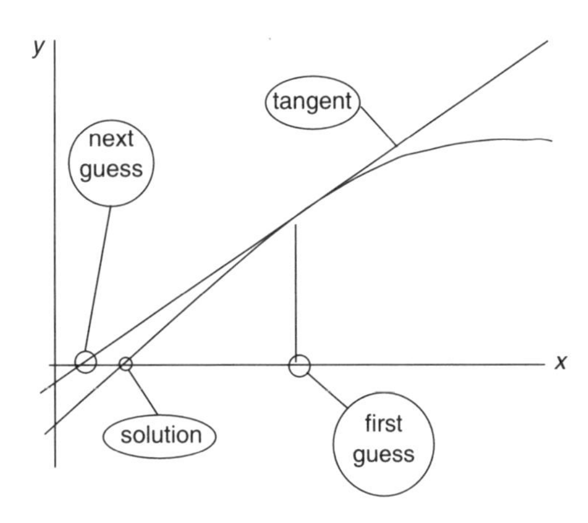

The tangent crosses the x axis, at a point closer to the solution than the first guess! We use this improved guess to repeat the process to continually improve on the guesses.

To use your CAS to find the equation of the tangent, firstly **DIFFERENTIATE** the function and **SUBSTITUTE** your x co-ordinate into this derivative, this will give you m the gradient of the tangent at x_1. Use the straight line equation $y - y_1 = m(x - x_1)$ to **INPUT** the equation of the tangent: x_1 is your x co-ordinate; y_1 is the corresponding y co-ordinate, found by **SUBSTITUTING** x_1 into your original function. Now **SOLVE** this expression for y. You now have the equation of the tangent to your function at x_1.

The tangent you have found is of the form $y = mx + c$ and $y = 0$ where the tangent crosses the x axis. Solving $mx + c = 0$ for x will give the next guess! Now find the equation of the tangent at this next guess and find where it crosses the x axis. Repeat this process until you are happy with the solution!

Be careful with 7(a) as the number e will differ from the variable e. Check with your teacher or CAS manual to see how you input the number e.

Teaching notes

Solutions

(2) The derivative of $x\sin x - 1$ is $\sin x + x\cos x$, substituting $x = 1$ into this expression gives $\sin(1) + \cos(1)$, which is approximately 1.38177. The y value at $x = 1$ is $\sin(1) - 1$, hence the equation of the tangent can be written as
$y - (\sin(1) - 1) = (\sin(1) + \cos(1))(x - 1)$
Solving for y and approximating gives
$y = 1.38177x - 1.54030$

(3)

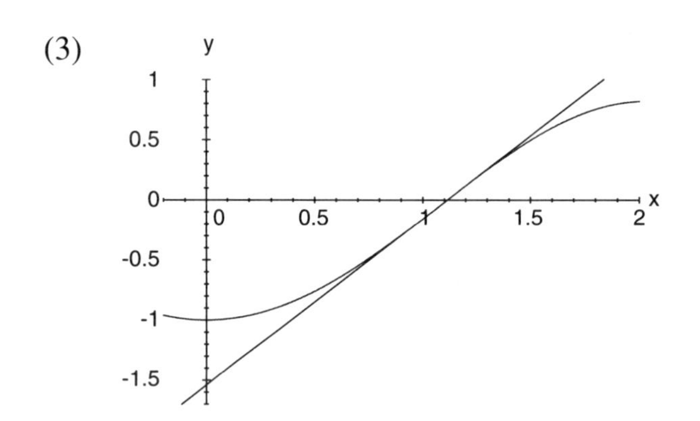

(4) Putting $y = 0$ and solving $1.38177x - 1.54030 = 0$ we obtain $x = 1.11472$, which is closer to the root than $x = 1$.

(5) We now repeat the above process to find the equation of the tangent of $y = x\sin x - 1$ at the point $x = 1.11472$. This is $y = 1.38874x - 1.54727$ and again putting $y = 0$ and solving we obtain $x = 1.11415$.

(6) We can see from this approximation that there appears to be some convergence. If we continue the process, working to 6 decimal places, there is no difference in the approximations from here on.

7(a) Be careful here, the number e will differ from the variable e. Check your manual to see how you input the number e.

Answer $x = -0.578713$

7(b) Answer $x = 1.52821$

The above method for solving equations numerically is called the Newton-Raphson method and is an iterative method for solving equations. It can be shown that the Newton-Raphson method for the solution of the equation $f(x) = 0$ can be written as

$$x_{n+1} = x_n - \frac{f(x_n)}{\frac{d}{dx}f(x_n)}.$$

If the above expression is constructed and iterated from a given starting value, the Newton-Raphson process will be automated. Different CASs have their own unique way of performing the iterative process. More able and confident students may wish to get their CAS to iterate the Newton-Raphson process. As an example with DERIVE the Newton-Raphson method for the solution of $x\sin x - 1 = 0$, starting at $x = 1$ could be calculated in the following way.

#1: $x\sin(x) - 1$

#2: $\frac{d}{dx}(x\sin(x) - 1)$

#3: $x - \frac{x\sin(x) - 1}{\frac{d}{dx}(x\sin(x) - 1)}$

#4: $\text{ITERATES}\left[x - \frac{x\sin(x) - 1}{\frac{d}{dx}(x\sin(x) - 1)}, x, 1, 5\right]$

#5: [1,1.11472,1.11415,1.11415,1.11415]

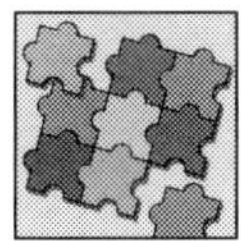

Extensions

(8) This question is an example of how an inappropriate starting value can lead to problems. It will take 48 applications of the Newton-Raphson method to get convergence to 6 decimal places. The Newton-Raphson method generally fails if the starting point is near a turning point. The shallow gradient makes the tangent cross the x axis a long way from the root. In some cases the tangent may cross the x axis in a region where the function is undefined, e.g. $\ln(x)$ is undefined for $x<0$, hence the next attempt to find a tangent is impossible in terms of real numbers.

Appendices

Derive (version 3)

Derive will run on almost any PC, as it requires only 512 KB of memory. It is mainly menu driven, although it can respond to commands, a great deal can be done through using the menu system alone. It also generates 2D plots and has some 3D graph capabilities, and comes with a set of on-line help files. *Derive for Windows* is also available.

APPROXIMATE

#1: $\frac{\pi}{4}$

#2: 0.785398

press X for approXimate

EXPAND

#1: $x\cdot(x + 1)^2\cdot(x - 1)$

#2: $x^4 + x^3 - x^2 - x$

press E for Expand

FACTORIZE

#2: $x^4 + x^3 - x^2 - x$

#3: $x\cdot(x - 1)\cdot(x + 1)^2$

press F for Factorize

DIFFERENTIATE

#1: $x\cdot\text{SIN}(x)$

#2: $\frac{d}{dx}(x\cdot\text{SIN}(x))$

#3: $x\cdot\text{COS}(x) + \text{SIN}(x)$

press C for Calculus, D for Differentiate and S for Simplify

PLOT

#1: x^2

press P for the Plot menu, then P again to draw the curve

SCALE/ZOOM

to go from first to second graph either press S for Scale or R for Range. press Z for Zoom and then select either (or both) *x* or *y*-axis and direction (zoom IN or zoom OUT).

SOLVE

#1: $x^3 - 2\cdot x - 4$

#2: $x = 2$

#3: $x = -1 + \hat{\imath}$

#4: $x = -1 - \hat{\imath}$

press L for soLve

INTEGRATE

#1: $x\cdot\text{SIN}(x)$

#2: $\int x\cdot\text{SIN}(x)\,dx$

#3: $\text{SIN}(x) - x\cdot\text{COS}(x)$

press C for Calculus, I for Integrate and S for Simplify

LIMIT

#1: $\frac{2\cdot x + 3}{3\cdot x + 4}$

#2: $\lim_{x\to\infty} \frac{2\cdot x + 3}{3\cdot x + 4}$

#3: $\frac{2}{3}$

press C for Calculus, L for Limit, 'inf' and S for Simplify

SUM

#1: i^{-2}

#2: $\sum_{i=1}^{\infty} i^{-2}$

#3: $\frac{\pi^2}{6}$

press C for Calculus, S for Sum, 'inf' and S for Simplify

SUBSTITUTE

#1: $x^n - 1$

#2: $x^7 - 1$

highlight the expression to be substituted, press M for Manage, then S for Substitute and then follow the prompts.

SIMPLIFY

see **Differentiate**, **Integrate**, **Limit** and **Sum** examples

MathPlus/Theorist (version 2)

MathPlus (called Theorist in the U.S.) is available for both PC/Windows and Macintosh. Expressions can be generated using palettes for both variables and functions, and a number of manipulations can be carried out directly by clicking and dragging with a mouse. It also generates 2D and 3D plots, which can be manipulated directly.

APPROXIMATE

○ $\frac{\pi}{4}$

△ $\frac{\pi}{4} = 0.785398$

Select MANIPULATE - Calculate. To alter the accuracy, select PREFS - Display Precision

EXPAND

○ $x\ (x+1)^2 (x-1)$

△ $x\ (x+1)^2 (x-1) = x^4 + x^3 - x^2 - x$

Select MANIPULATE - Expand. This may be used more than once. Miniexpand only expands the outer layer.

FACTORIZE

○ $x^4 + x^3 - x^2 - x$

△ $x^4 + x^3 - x^2 - x = (x+1)^2 (x-1) x$

Select MANIPULATE - Factorize. MANIPULATE - Collect gives common factors.

DIFFERENTIATE

○ $\frac{\partial}{\partial x}(x\ \sin[x])$

△ $\frac{\partial}{\partial x}(x\ \sin[x]) = x\ \cos(x) + \sin(x)$

Select MANIPULATE - Simplify once the expression has been entered using the function palette

SOLVE

○ $x^2 - 2x - 4 = 0$

○ ○ $\frac{1}{2}(\sqrt{20} + 2) = x$

○ ○ $\frac{1}{2}(-\sqrt{20} + 2) = x$

Highlight both occurrences of the x and option-drag to the right hand side of the equation.

PLOT

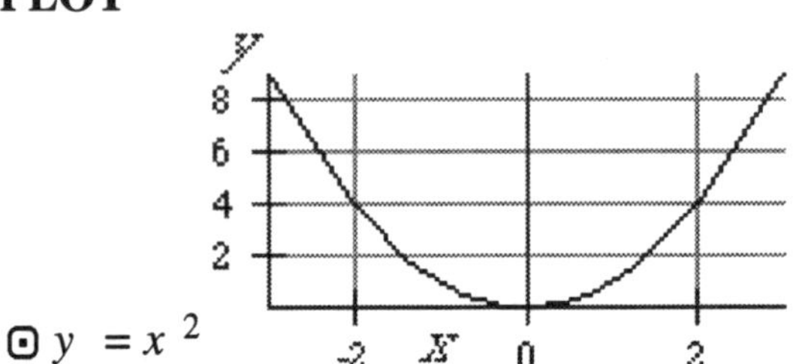

⊡ $y = x^2$

Highlight function, then select GRAPH - y = f(x) - Linear.

INTEGRATE

△ $\int x^2 \sin(x)\, d\, x =$

$-x^2 \cos(x) + 2(\cos[x] + x\ \sin[x])$

Select MANIPULATE - Int. by Parts and then MANIPULATE - Simplify several times.

SCALE/ZOOM

Three tools are available: The first two are for zooming in and out, while the third allows numerical change of the domain and range.

LIMIT

MathPlus will not calculate limits symbolically. Estimates can be made graphically or numerically.

SIMPLIFY

See Differentiate, Integrate for examples. Part of an expression can be selected to be simplified first. Collect and Factor can also help to simplify expressions.

SUBSTITUTE

○ $n = 7$ drag ○ $n = 7$

○ $x^n - 1$ ○ $x^n - 1$

△ $x^n - 1 = x^7 - 1$

Highlight the substitution equation and option-drag over the substitution variable

SUM

△ $\sum_{k=1}^{10} k^{-2} = 1.5498$

The sum operator is chosen (see **INPUT**). Then select MANIPULATE - Calculate.

Maple V (release 4)

Maple V is a powerful computer algebra system which runs on a number of platforms, including PC/Windows and Macintosh. It is command driven and programmable, although some commands can be menu-driven on some versions. A wide range of 2D and 3D plot options are also available, and it has a built-in help system.

APPROXIMATE

```
> evalf(Pi/4);
```

$$.7853981635$$

EXPAND

```
> expand(x * (x + 1)^2 * (x - 1));
```

$$x^4 + x^3 - x^2 - x$$

FACTORIZE

```
> factor(x^4 + x^3 - x^2 - x);
```

$$x(x+1)^2(x-1)$$

DIFFERENTIATE

```
> diff(x * sin(x), x);
```

$$\sin(x) + x\cos(x)$$

SOLVE

```
> solve(x^2 - 2 * x - 4 = 0);
```

$$\sqrt{5}+1, 1-\sqrt{5}$$

```
> solve({y=x+2,y=2-x-x^2});
```

$$\{x=-2, y=0\},$$
$$\{x=0, y=2\}$$

PLOT

```
> plot((x + 2) * (1 - x),x=-3...3);
```

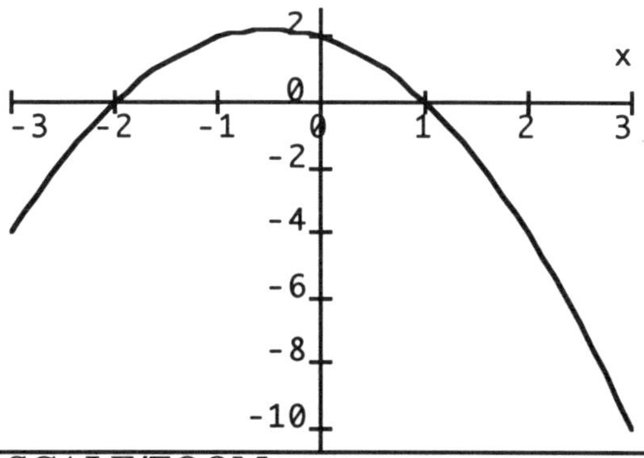

INTEGRATE

```
> integrate(x * sin(x),x);
```

$$\sin(x) - x\cos(x)$$

SCALE/ZOOM

To scale or zoom, plot again with a different range and/or domain; e.g.

```
> plot(sin(x),x=-3...3);
> plot(sin(x),x=0..1,y=0..1);
```

LIMIT

```
> limit(exp(x), x=-infinity);
```

$$0$$

SIMPLIFY

```
> simplify(exp(a+ln(b*exp(c))));
```

$$b\mathbf{e}^{(a+c)}$$

SUBSTITUTE

```
> subs(n=7, x^n-1);
```

$$x^7 - 1$$

SUM

```
> sum(1/k^2,k=1..10);
```

$$\frac{1968329}{1270080}$$

TI-92

The Texas Instruments TI-92 is a graphic calculator with a number of built in features including a computer algebra system. The computer algebra system was developed jointly by TI and the makers of Derive and is mainly menu driven - although commands can also be typed in directly. It also generates 2D plots and has some 3D plot capabilities.

APPROXIMATE

F1 | F2 Algebra | F3 Calc | F4 Other | F5 PrgmIO | F6 Clear a-z...

■ $\frac{\pi}{4}$ $\quad\quad$ $\frac{\pi}{4}$

■ approx$\left(\frac{\pi}{4}\right)$ $\quad\quad$ .785398163397

approx(π/4)

MAIN RAD AUTO FUNC 2/30

Select F2, select *Approx*, enter *expression*)

EXPAND

■ expand$(x\cdot(x-1)\cdot(x+1)^2)$ $\quad\quad$ $x^4+x^3-x^2-x$

expand(x*(x-1)*(x+1)^2)

Select F2, select *Expand*, enter *expression*.)

FACTORIZE

■ factor$(x^4+x^3-x^2-x)$ $\quad\quad$ $x\cdot(x-1)\cdot(x+1)^2$

factor(x^4+x^3-x^2-x)

Select F2, select *Factorize*, enter *expression*.)

DIFFERENTIATE

F1 | F2 Algebra | F3 Calc | F4 Other | F5 PrgmIO | F6 Clear a-z...

■ $\frac{d}{dx}(x\cdot\sin(x))$ $\quad\quad$ $x\cdot\cos(x)+\sin(x)$

d(x*sin(x),x)

MAIN RAD AUTO FUNC 1/30

Select F3 menu, select *Differentiate* and enter *expression* , *variable* ,)

SOLVE

■ cSolve$(x^3-1=0,x)$

$x=-1/2+\frac{\sqrt{3}}{2}\cdot i$ or $x=-1/2-\frac{\sqrt{3}}{2}\cdot i$ or $x=1$

cSolve(x^3-1=0,x)

Select F2 menu and select *Solve* or *cSolve* (which allows complex solutions). Correct syntax [*equation* , *variable*] is required.

PLOT

F1 | F2 Zoom | F3 Trace | F4 ReGraph | F5 Math | F6 Draw | F7

PLOTS
$y1=x^2$
y2=
y3=
y4=
y5=
y6=

Select *Y*= to enter function, select *Graph*.

INTEGRATE

■ $\int(x\cdot\sin(x))dx$ $\quad\quad$ $-(x\cdot\cos(x))+\sin(x)$

∫(x*sin(x),x)

Select F3, select *Integrate* and enter *expression* , *variable*).

SCALE/ZOOM

F1 | F2 Zoom | F3 Trace | F4 ReGraph | F5 Math | F6 Draw | F7

1:ZoomBox
2:ZoomIn
3:ZoomOut
4:ZoomDec
5:ZoomSqr
6:ZoomStd
7:ZoomTrig
8:ZoomInt
9:ZoomData
A:ZoomFit
B:Memory
C:SetFactors...

TYPE OR USE ←→↑↓ + [ENTER]=OK AND [ESC]=CANCEL

Window allows ranges to be set. The *Zoom* options are illustrated above.

LIMIT

■ $\lim_{x\to\infty}\left(\frac{2\cdot x+3}{3\cdot x+4}\right)$ $\quad\quad$ 2/3

limit((2*x+3)/(3*x+4),x,∞)

Select F3, select *Limit*, enter *expression, variable, limit*).

SIMPLIFY

This does not exist as a separate command but can be performed through *Expand*, *Factor*, etc. There is not need to ask the machine to simplify derivatives and integrals as, for example, is the case with DERIVE.

SUBSTITUTE

F1 | F2 Algebra | F3 Calc | F4 Other | F5 PrgmIO | F6 Clear a-z...

■ $x^n-1|n=7$ $\quad\quad$ x^7-1

x^n-1|n=7

MAIN RAD AUTO FUNC 1/30

Immediately after the expression use the 'with operator', | (2nd *k*), followed by the substitution

SUM

■ $\sum_{i=1}^{\infty}(i^{-2})$ $\quad\quad$ $\frac{\pi^2}{6}$

Σ(i^-2,i,1,∞)

Select F3 menu and select *Sum*. Correct syntax [*expression, variable, lower limit, upper limit*] is required.

Mathematica (version 2)

Mathematica is a powerful computer algebra system which runs on PC/Windows, Macintosh and other platforms. Like Maple, it is command driven and has a built-in programming language. A wide range of 2D and 3D plot options are also available, together with on-line help.

APPROXIMATE

```
N[Pi/4,10]

        0.7853981634
```

EXPAND

```
Expand[x((x + 1)^2)(x - 1)]
                 2    3    4
           -x - x  + x  + x
```

FACTORIZE

```
Factor[x^4 + x^3 - x^2 - x]
                    2
       (-1 + x) x (1 + x)
```

DIFFERENTIATE

```
D[x Sin[x], x]

           x Cos[x] + Sin[x]
```

SOLVE

```
Solve[x^2 - 2x - 4 == 0, x]

         2 - 2 Sqrt[5]          2 + 2 Sqrt[5]
{{x -> -------------},{x -> -------------}}
               2                      2

Solve[{y == x+2,
            y == 2-x-x^2},{x,y}]

{{y -> 0, x -> -2}, {y -> 2, x -> 0}}
```

PLOT

```
Plot[(x + 2)(1 - x), {x, -3, 3}]
```

INTEGRATE

```
Integrate[x Sin[x], x]

        -(x Cos[x]) + Sin[x]
```

SCALE/ZOOM

To scale or zoom, change the range and/or domain by using the option `PlotRange`; e.g.

```
PlotRange->{ymin, ymax}
PlotRange->{{xmin, xmax},
                     {ymin, ymax}}
```

LIMIT

```
Limit[E^x, x->-Infinity]

        0
```

SIMPLIFY

```
Simplify[Sin[x]^4-cos[x]^4]

           -Cos[2 x]
```

SUBSTITUTE

```
x^n-1 /. n->7

              7
        -1 + x
```

SUM

```
Sum[1/k^2, {k, 1, 10}]

           1968329
           -------
           1270080
```

Macsyma (version 2)

Macsyma is a powerful CAS with detailed on-line help - available for PC/Windows and UNIX systems. Like Maple and Mathematica, Macsyma is command driven and programmable, but also supports menu-driven templates for most commands. A wide range of 2D and 3D plot options are also available.

APPROXIMATE

(c1) $\mathrm{dfloat}\left(\frac{\pi}{4}\right)$

(d1) 0.78539816339744

EXPAND

(c1) $x*(x+1)\wedge 2*(x-1), \mathrm{expand}$

(d1) $x^4 + x^3 - x^2 - x$

FACTORIZE

(c1) $x\wedge 4 + x\wedge 3 - x\wedge 2 - x, \mathrm{factor}$

(d1) $(x-1)\,x\,(x+1)^2$

DIFFERENTIATE

(c1) $\frac{d}{dx}(x\sin(x))$

(d1) $\sin(x) + x\cos(x)$

SOLVE

(c1) $\mathrm{solve}\left(\left[2x - y, 3x + y^2\right], [x, y]\right)$

(d1) $\left[[x = 0, y = 0], \left[x = -\frac{3}{4}, y = -\frac{3}{2}\right]\right]$

INTEGRATE

(c1) $\int x\sin(x)\,dx$

(d1) $\sin(x) - x\cos(x)$

PLOT

(c1) plot(x^2,x,−1,1),equalscale

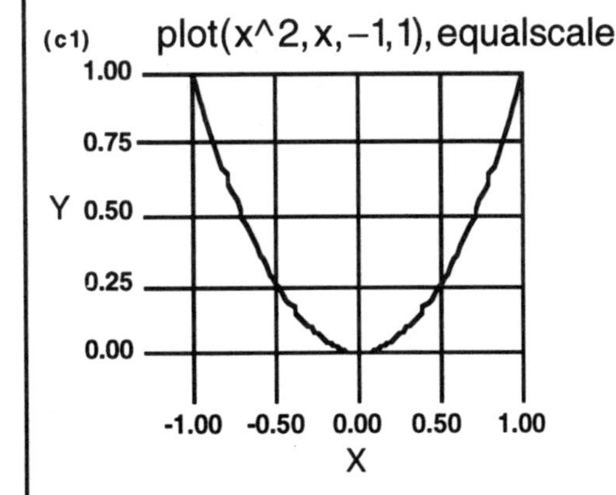

SCALE/ZOOM

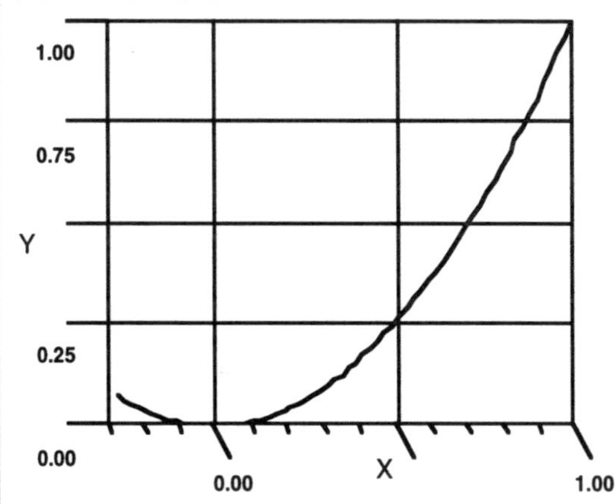

LIMIT

(c1) 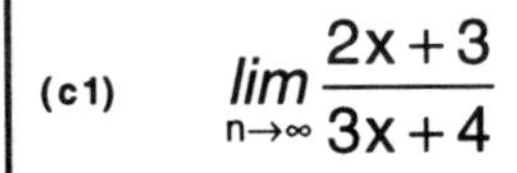 $\lim_{n\to\infty} \frac{2x+3}{3x+4}$

(d1) $\frac{2}{3}$

SIMPLIFY

(c1) $\exp(a + \log(b * \exp(c))), \mathrm{logexpand}$

(d1) $b\,e^{c+a}$

SUBSTITUTE

(c1) $(x^n - 1, n = 7)$

(d1) $x^7 - 1$

SUM

(c1) $\sum_{i=1}^{10} \frac{1}{i^2}$

(d1) $\frac{1968329}{1270080}$